上海大学出版社

2005年上海大学博士学位论文 41

U0358902

非线性全局优化的变换函数方法

- 作者：王　薇
- 专业：运筹学与控制论
- 导师：张连生

A Dissertation Submitted to Shanghai University for the
Degree of Doctor in Science（2005）

Transformation Function Methods for
Nonlinearly Global Optimization

Ph. D. Candidate: WANG　Wei
Supervisor: ZHANG Liansheng
Major: Operations Research & Cybernetics

Shanghai University Prees
• Shanghai •

摘　　要

最优化是一门应用相当广泛的学科,它讨论决策问题的最佳选择,构造寻求最佳解的计算方法并研究这些方法的理论性质及实际计算表现. 由于社会的进步和科学技术的发展,最优化问题广泛见于经济计划、工程设计、生产管理、交通运输、国防军事等重要领域,因此受到高度重视. 伴随着计算机的高速发展和最优化工作者的努力,最优化的理论分析和计算方法得到了极大提高. 本论文主要工作就是讨论、研究了非线性最优化问题的几个算法及理论分析.

本文包含五章内容. 第一章简述了目前国内外几种主要的全局最优化问题和算法及本论文所要用到的非线性规划的一些基本概念及性质. 后面四章由四篇基本独立的文章组成.

第二章和第三章主要讨论求解无约束全局最优化问题的变换函数法. 求解一般函数的全局最优解问题是热点课题之一. 对全局问题有两个困难需要解决. 一是如何从一个局部极小解出发找到更好的局部解,另一个是全局最优解的判定问题. 打洞函数法和填充函数法是解决第一个困难的实用方法. 它们的共同点是如果已经找到了一个局部极小 x_1^*,但它不是全局最小,我们可以在 x_1^* 处构造一个辅助函数——打洞函数或填充函数使迭代点列离开 x_1^* 所在的谷域,找到更好的点 x' (即 x' 处的函数值比 x^* 处的函数值更小). 然后以 x' 为起点找出更优的局部极小点. 第二章定义了两类变换函数,在适当的条件下证明了它们兼具打洞函数和填充函数的特点和性质,即填充函数法和打洞函数法两种方法在某种意义下是可以统一的，因此可称其为 T–F 函数. 第三章给出了几个简单、易于计算且函数性态较好的变换函数，它们同样兼具打洞函数和填充函数的特点和性质. 文章证明了第二、三章定义的变换函数的主要性质：在 $f(x)$ 的值比当前局部极小值 $f(x_1^*)$ 大的水平集上变换函数没有极小点或稳定点；在比当前局部极小值小的水平集上变换函数一定有极小值点. 当然这两章也给出了数值试验结果.

第四章将用于无约束全局最优问题的思想方法拓广到求解带有约束的非线性规划问题的全局最优问题. 首先，对于求解带有线性约束的非线性规划问题的局部解，已有很多算法. 但这些方法考虑的都是局部问题，所以要得到全局解必需要有"凸"性条件. 本章去掉目标函数的凸性要求，在比较宽松的假设下给出了一个算法，证明了算法的一些理论性质. 然后，对于带有非线性约束的一般形式的全局优化问题给出了另一个算法，此算法对目标函数和约束函数都没有凸性要求，同样在较宽松的假设下证明了算法和变换函数的主要理论性质.

第五章提出一个用于求解带不等式约束非线性规划问题的修正共轭梯度投影方法. 由于拟牛顿法具有较快的收敛性，所以它仍然是非线性规划领域的重要方法之一. 相比于其他约束拟牛顿方法，本章的方法有两个优点：一是不用严格互补条件而证明了算法的全局收敛性和局部超线性收敛性，二是每步迭代只需计算一次投影矩阵从而减少了计算量. 搜索方向是显式的.

关键词 非线性规划，全局最优解，变换函数，填充和打洞函数，梯度投影，超线性收敛

Abstract

The optimization is a widely used discipline, which discusses the characters of optimal choice on decision problems and constructs computing approaches to find the optimal solution. Due to the advancement of society and the development of science and technology, the optimization problems are often discovered in the field of economic planning administration, engineering design, production management, traffic transportation, national defence and so on. They are so important that meet with much recognition. With the speedy development of computer and the hard work of scientists, the theoretic analysis and computational methods on optimization have been highly improved .

This paper mainly consists of five chapters.

In the first chapter, some mainly methods for global optimization problems are briefly presented. And several basis concepts and characters on generally nonlinear programming are introduced.

In the second and third chapter the transformation functions for unconstrained global optimal problems are mainly discussed. To find the effective methods for finding the global optimal solutions of a general multi-minimizers function is one of the hot topics. There two difficulties in global optimization. One is how to leave from a local minimizer to a smaller one and the other is how to judge that the current minimizer is global. The tunnelling function proposed by Levy and Montalvo (1985) and filled function algorithms introduced by Ge and Qin (1987) are two well-known and practical methods for settling the first difficulty. They have common character. If a local minimizer x_1^* has been found, we can make a auxiliary function, such as tunnelling function or filled function, such that iterative sequential points leave the valley in which x_1^* lies to find a better point x' in the lower valley (i. e. $f(x') \leqslant f(x_1^*)$). Then let x' be a new initial point to search for a better minimizer. In second chapter two classes of transformation functions for global optimization are defined and it is proved theoretically and computationally that they possess the both characters of tunnelling functions and filled functions under some general assumptions. In third chapter some easy and computable transformation functions are presented. They have the both characters of tunnelling functions and filled functions as well. We proved the main characters of transformation functions, that is, the transformation functions have no any minimizer or stationary point on the region $\{x: f(x) \geqslant f(x_1^*)\}$ and have at least one minimizer on the region $\{x: f(x) < f(x_1^*)\}$ if $\{x: f(x) < f(x_1^*)\} \neq \varnothing$. Certainly, the numerical results are listed in these two chapters.

In chapter four, the idea for unconstrained global

optimization is extended to nonlinear global problems with constraints. First, there exist many effective methods for local minimizers of nonlinear programming. Because these methods are used for local problems, there should be "convex" condition in order to get a global minimizer. Now we get rid of the convexity and propose an algorithm for linear constrained global optimization with some proper assumptions. The properties of algorithm have been proved theoretically. Then, another algorithm for general global optimal problem with nonlinear constraints is presented. The convexity on objective function and constraints is still not asked in this algorithm. In the same way, the theoretical properties of algorithm and transformation function are proved under some gentle assumptions.

In fifth chapter, a revised conjugate projection gradient method for nonlinear inequality constrained optimization problems is proposed. The search direction in the method is the combination of the conjugate projection gradient and the quasi - Newton direction. Owing to the fast convergence, the constrained quasi - Newton methods are still the important topics. There are two merits in our method. The one is the amount of computation is fewer since the gradient matrix only needs to be computed once in each iteration. The other is that the algorithm is super - linearly convergent without strict complementary condition under some mild assumptions.

Key words　Nonlinear programming，　Global minimizer，
Transformation function，　Tunnelling and filled function，
Gradient projection，　Suplinear convergence

目　录

第一章　全局最优化问题
概述及预备知识

§1.1　最优化问题模型

　　最优化是应用数学领域的重要组成部分,最优化问题广泛见于金融经济模型,工程设计,生产管理,网络交通,农业预测,国防军事等重要领域,因此受到高度重视. 最优化包含很多分枝,如线性规划,非线性规划,组合优化,多目标规划等等. 本论文主要讨论非线性规划问题. 非线性规划被用来识别和计算多个变量的非线性函数的最优解. 如果这些变量受到一些条件的限制时,称其为约束最优化问题. 如果变量可以自由变动不受约束的限制,则称其为无约束最优化问题.

　　有约束最优化问题可表示为

$$\min f(x)$$

$$s.t. \ g_j(x) \leqslant 0, j \in L \qquad (1.1.1)$$

$$g_j(x) = 0, j \in M$$

这里 $x \in \Omega \subset R^n$, $L = \{1, 2, \cdots, m\}$ 和 $M = \{m+1, \cdots, m+p\}$ 为指标集. $f(x)$, $g_j(x): R^n \to R$,称 $f(x)$ 为目标函数, $g_j(x)$ 为约束函数; $f(x)$ 和 $g_j(x)(j \in L \bigcup M)$ 至少一个是非线性的.

　　无约束最优化问题可表示为

$$\min_{x \in R^n} f(x) \qquad (1.1.2)$$

问题(1.1.1)的可行域 $X = \{x \mid g_j(x) \leqslant 0, j \in L; g_j(x) = 0, j \in M; x \in \Omega\}$;问题(1.1.2) 的可行域 $X = \{x \mid x \in \Omega \subset R^n\}$.

为方便统称问题(1.1.1)和问题(1.1.2)为**原问题**.

定义 1.1.1 至少有一个 $x_G^* \in X$ 使得 $\forall x \in X$ 有 $f(x) \geqslant f(x_G^*)$,或证明这种点不存在,这样的问题称为全局极小化问题. 若 x_G^* 存在则称其为原问题的全局极小点或全局解. 对应的函数值 $f(x_G^*)$ 称为 $f(x)$ 在 X 上的全局最优值.

令 $\|\cdot\|$ 代表 R^n 中的 Euclidean 范数. 则有

定义 1.1.2 若存在 $x^* \in X$ 的邻域 $N(x^*) = \{x \mid \|x - x^*\| < \delta, \delta > 0\}$ 使得 $\forall x \in N(x^*) \bigcap X$ 有 $f(x) \geqslant f(x^*)$,则称 x^* 为原问题的局部解(或局部极小点).

关于算法的搜索迭代方向有下面的定义.

定义 1.1.3 设 $x^0 \in X$. 若有方向 $d \in R^n$ 且 $d \neq 0$ 使得 $d^T \nabla f(x^0) < 0$,则称 d 为 x^0 处的下降方向;若有 $\alpha > 0$ 使得 $x^0 + \alpha d \in X$,则称 d 为 x^0 处的可行方向.

定义 1.1.4 对于函数 $f(x)$,若有函数 $\phi(t)$ 使得 $\phi[f(x)]$ 改变了 $f(x)$ 的函数形态和性质,则称 $\phi(t)$ 为 $f(x)$ 的变换函数.

§1.2 全局最优化问题及算法概述

由于科学和工程等领域所需要解决的问题越来越依赖于问题的全局解,促使过去三、四十年里,关于全局最优化的新理论和算法层出不穷. 由于很可能在一个全局优化问题里有许多局部极小点,因此全局优化问题不能简单地用通常意义上的求解目标函数局部极小的方法. 这些多极小值点的函数一般会导致两个比较困难的问题:第一个是怎样离开一个局部极小值点到一个更优的极小点;第二个是怎样判断当前的极小值点是否为全局最优解. 现在有很多方法研究解决第一个问题,而第二个问题——全局收敛条件的研究仍然比较困难,没有突破性的工作.

全局优化问题的特点导致了其算法不同于经典传统方法. 一般地讲,求解全局问题的方法可分为两类:一类是确定性算法,一类是随机算法.

确定性算法是利用问题的解析性质产生一个确定性的有限或无限的点序列使其收敛于全局最优解. 如 *D. C.* 规划算法、单调规划、分枝定界方法、填充函数法、打洞函数法和积分水平集方法[89]等. 凸性、单调性、稠密性、等度连续性、李普希兹常数、高阶导数、一致性、水平集等通常称为全局性的解析性质.

随机算法利用概率机制而非确定性的点列来描述迭代过程. Monto - Carlo 方法、遗传算法、模拟退火算法是常用的随机算法. 这些方法以及禁忌搜索、人工神经网络等被称为现代优化计算方法. 现代优化算法通常是通过模拟生物进化、人工智能、数学与物理科学、神经系统和统计力学等概念,以直观为基础构造的,此类算法我们亦称为启发式算法. 很多实际问题的目标函数解析性态差甚至没有解析式,传统的建立在梯度计算基础上的非线性规划算法受到限制. 随机方法(如遗传算法)的并行性,广泛的可适用性(如对目标函数的性态无特殊要求,特别可以没有明确的表达式)和较强的鲁棒性,简明性与全局优化性能等优点受到关注.

下面简单概述几个全局优化的确定及随机算法.

§1.2.1 *D. C.* 规划

当目标函数和约束函数均为凸函数时,此类凸规划问题局部解就是全局解,很多非常有效的经典算法可以解决这类问题. 但当目标函数和约束函数至少有一个不是凸函数时,问题就变得相当复杂了,至今没一个非常有效的解决方法. 然而,在实际问题中经常会遇到下述问题:

$$\min f(x) = c_{0,1}(x) - c_{0,2}(x)$$

$$s.t. \ f_i(x) = c_{i,1}(x) - c_{i,2}(x) \leqslant 0 \quad (i = 1, 2, \cdots, m)$$

$$x \in X \subset R^n \tag{1.2.1}$$

其中 X 为 R^n 中的紧凸集，且 $c_{i,1}(x)$，$c_{i,2}(x)$ $(i=0,1,\cdots,m)$ 均为凸函数. 通常我们称表示为两个凸函数差的函数为 $D.C.$ 函数，上述问题(1.2.1) 称为 $D.C.$ 规划问题.

下面简单地介绍一下 $D.C.$ 规划中的一些性质：

1) 任意一个定义在 R^n 的紧凸集上的二次可微函数(特别是多项式函数)为 $D.C.$ 函数.

2) 任何闭集 $S \subset R^n$ 都能表示成 $D.C.$ 不等式的解集：$S = \{x \in R^n \mid g_s(x) - \|x\|^2 \leqslant 0\}$ 其中 $g_s(x)$ 为 R^n 上的一个连续凸函数.

3) 设 $f_1(x),\cdots,f_m(x)$ 是 $D.C.$ 函数，则函数 $\sum_{i=1}^m \alpha_i f_i(x)$，$(\alpha_i \in R)$，$\max\limits_{1\leqslant i\leqslant m} f_i(x)$ 和 $\min\limits_{1\leqslant i\leqslant m} f_i(x)$ 均为 $D.C.$ 函数.

利用上述性质，每一连续规划问题可以化成一个带线性目标函数及不多于一个凸和一个反凸约束的 $D.C.$ 规划问题. 详细内容可见文献[32]. 但这只是原则上说任何一个连续规划问题可以化为一个 $D.C.$ 规划问题，如何把一个具体的规划问题转化为一个等价的 $D.C.$ 规划问题还是比较困难的.

一个非常值得研究的全局优化问题之一为带有线性约束的，目标函数为凹的最小化问题，亦称带线性约束的凹规划问题，其可表述为一定义在多面体 $D \subset R^n$ 上的凹函数的全局极小问题：

$$\min\{c(x) \mid x \in D\}, \tag{1.2.2}$$
$$D = \{x \in R^n \mid a_i^T x \leqslant b_i, i=1,\cdots,m\}$$

20 世纪 70 年代，对此问题作了非常详尽的研究. 30 多年来在对这个问题研究过程中的很多思想被进一步应用到了更一般的 $D.C.$ 规划中. 1994 年以前这方面的主要成果可参见文献[33].

最早应用凸规划的外逼近方法被成功地应用于凹规划，以后又被成功地应用于反凸规划、$D.C.$ 规划及单调规划问题. 由于定义在多面体上的凹函数的最小值在顶点上达到,所以基于上述性质可以通过构造一簇多面体序列 $P_1 \supset P_2 \supset \cdots \supset D$ 用问题 $\min\{c(x) \mid x \in P_k\}$ 的解逼近于 $\min\{c(x) \mid x \in D\}$ 的解,其中 P_{k+1} 是由多面体 P_k 添加线性约束后所构成的. 通过文献[34]中的一个有效方法,从 P_k 导出相应的顶点集 V_k,然后计算 $x^k \in \text{argmin}\{c(x) \mid x \in V_k\} = \text{argmin}\{c(x) \mid x \in P_k\}$,并且得到序列 $\{x^k\}$ 的聚点即问题的全局最优点.

我们可将外逼近方法表述如下:

步 0　构造一多面体 $P_1 \subset D$,计算它的顶点集 $V_1,k:=1$.

步 1　解问题 $\min\{c(x) \mid x \in P_k\}$ 的解 x^k,即 $x^k \in \text{argmin}\{c(x) \mid x \in V_k\}$,$\beta = c(x^k)$,$\beta \leqslant c^* = \min\{c(x) \mid x \in D\}$.

步 2　若 $I(x^k) = \{i \mid a_i^{\text{T}} x^k > b_i, i \in \{1, 2, \cdots, m\}\} = \varnothing$,即 $x^k \in D$,则算法终止;否则转步 3.

步 3　选择 $j \in I(x^k)$,令 $P_{k+1} = P_k \bigcap \{x \mid a_j^{\text{T}} x \leqslant b_j\}$,计算 $V(P_{k+1})$,转步 1.

外逼近方法的步骤非常简捷,它已被广泛用于凹规划、反凸规划和 $D.C.$ 规划及单调规划问题,详尽的结果可参见文献[33]. 外逼近方法的收敛性已由 Horst 等人在文献[34]给出.

定理 1.2.1　基于上述的外逼近方法,算法具有 m 步终止.

§1.2.2　单调规划

在经济、工程及其他一些领域中的大量数学模型通常都具有某些变量或所有变量的单调性的性质. 在最优设计的相当多数量的文章中,单调性在数值方法的研究中起着相当重要作用. 当单调性与凸性和反凸性结合在一起时,产生了乘积规划[32],C-规划[41]等等低维的非凸问题. 在过去的十年间,参数方法、对偶基的补偿方法对求解低维的上述问题的速度是相当快的.

下面我们考虑下述问题:

$$\min\{f(x) \mid g(x) \leqslant 1 \leqslant h(x), \ x \in R_+^n\} \qquad (1.2.3)$$

此处 $f(x)$，$g(x)$，$h(x)$ 都是单调增加的(即当 $0 \leqslant x \leqslant x'$，$f(x') \geqslant f(x)$，称 $f(x)$ 为增加的). 由考虑的抽象凸性,在某种特殊假设下,此类问题可由广义外逼近方法来处理. 然而,纯单调结构的最重要的优点在于利用全局信息,通过在可行域的限制区域上的极限的全局搜索,可以简化问题. 事实上,当(1.2.3)的目标函数是单调增加时,若 z 为可行域已知的可行点,因在 $z + R_+^n$ 没更好的可行解,故 z 在 $z + R_+^n$ 上是不起作用的. 类似地,当函数 $g(x)$(相应的 $h(x)$)是单调增加时, z 为在 $z + R_+^n$ 上对于约束 $g(x) \leqslant 1$(相应的 $h(x) \geqslant 1$)为不可行点, 则整个 $z + R_+^n$ 可以不予考虑. 基于上述观察,外逼近或分枝定界等有效方法来可以用求解易处理的单调规划.

如果存在一单调增加函数 $g(x)$,使得 $G = \{x \in R_+^n \mid g(x) \leqslant 1\}$, 其形如 $G = \bigcup_{z \in Z} [0, z]$. 其中 $G = \bigcup_{z \in Z} [0, z]$ 为箱子簇 $[0, z]$ 的和集, $z \in Z$,则称 G 是正规的. 当 Z 为有限集时,正规集称为多胞块. 正像紧凸集是一簇多胞体的交集,紧正规集是一簇多胞块的交集. 由此, 单调系统的解集结构的特征可以被建立起来,用于单调不等式和单调优化问题的数值分析. 更重要的是,多胞块逼近方法可以推广到求解两个单调增加函数差的优化问题(亦称为 $D.I.$ 函数的规划问题, 比如问题(1.2.1)). 参见文献[64, 65]. 由于 n 个变量的多项式可表为两个正系数的多项式之差,即两个 R_+^n 上的单调增加函数之差,由 Weierstrass 定理可知,在 $[0, b] = \{x \in R^n \mid 0 \leqslant x \leqslant b\}$ 上的 $D.I.$ 函数在 $[0, b]$ 上为稠密的,因此,$D.I.$ 最优化的适用范围包括多项式规划(特别是非凸二次规划)和各类全局和组合优化问题. 虽然如此, 但到目前为止,解 $D.I.$ 规划问题仍然是一个较困难的问题,在理论和算法上都没有 $D.C.$ 规划完备.

§1.2.3 分枝定界方法

在组合最优化中解全局最优的最普遍工具是应用分枝定界原

理,特别地在求解(1.2.2)时,将区域 D 划分成多面体子集,即划分成单纯形(单纯形剖分),划分成超长方体(超长方体剖分)或划分成锥体(锥剖分). 通常情况下,在小区域上易确定目标函数值的上下确界,从而逼近全局最优值. 该方法已广泛应用于凹规划,D.C. 规划及 Lipschitzian 规划.

首先用分枝定界方法求解下述全局优化问题:

$$\min_{x\in D} f(x)$$

其中 $f: R^n \to R$, $D \subset R^n$, D 为紧集,f 在 D 上连续.

下面介绍分枝定界方法的主要步骤.

步 1 选择初始可行域 M_0, $M_0 \supset D$, 把 M_0 划分为有限个子集 M_i, $i \in I$, I 是指标集. 划分要满足条件:

$$M_0 = \bigcup_{i\in I} M_i,$$

$$M_i \bigcap M_j = \partial M_i \bigcap \partial M_j, \quad \forall\, i, j \in I, i \neq j.$$

其中 ∂M_j 表示 M_j 的边界.

步 2 对每个子集 M_i 确定满足下面条件的上,下界 $\alpha(M_i)$, $\beta(M_i)$:

$$\beta(M_i) \leqslant \inf f(M_i \bigcap D) \leqslant \alpha(M_i).$$

再令 $\beta = \min\{\beta(M_i) \mid i \in I\}$, $\alpha = \max\{\alpha(M_i) \mid i \in I\}$, 则有

$$\beta \leqslant \min f(D) \leqslant \alpha.$$

步 3 若 $\alpha = \beta$ 或 $\alpha - \beta \leqslant \varepsilon$, ε 为充分小的正数,则算法终止. 否则转步 4.

步 4 选择适当的子集 M_i, 作更细的划分,转步 2.

分枝定界方法的实现主要是划分、选择和定界三个运算步骤. 划分、选择和定界的不同,产生相应不同的实现算法.

一般的,多面体或者凸多面体的划分的最简单形式是:单纯形,超长方体和多面锥. 从而划分集合 M_0 由它们所组成,并且一般都假

定细分是彻底的."细分是彻底的"的定义如下：

定义 1.2.1 多面体的一个细分是彻底的,如果由连续细分多面体所产生的每一个下降子列 $\{M_q\}$ 趋向于单点集;凸多面体的一个细分是彻底的,如果由连续细分凸多面体所产生的每一个下降子列趋向于一条射线.

Tuy 等在[62]中讨论了一大类单纯形的彻底的细分. 用得最多的是二等分,参见[28-30]. 其主要思想是：

设 M 是一个 n 维单纯形,它是由 $n+1$ 个仿射独立的顶点所组成的凸包. 设 $[v_M^r, v_M^s]$ 是 M 的最长的边,令 $v = \frac{1}{2}(v_M^r, v_M^s)$,用 v 分别代替顶点 v_M^r 和 v_M^s,这样产生二个单纯形,其体积之和等于 M 的体积. 这就是单纯形的二等分.

设 S 是一个内部含原点 O 的单纯形,(比如,令 S 的 $n+1$ 个顶点为 $v^i = e^i(i=1,2,\cdots,n)$, $v^{n+1}=-e$,其中 e^i 是第 i 个分量为 1 的 R^n 中的单位向量,$e = (1,\cdots,1)^T \in R^n$). 考虑 S 的 $n+1$ 个斜面 F_i,每一个 F_i 是 $n-1$ 维单纯形. 对每个 F_i,令 C_i 是一个凸锥,它的顶点是原点 O,n 条边是从 O 开始,通过 F_i 的 n 个顶点的射线. 于是 $\{C_i \mid i=1,2,\cdots,n+1\}$ 是 R^n 的一个锥划分. 如果对初始的斜面进行细分,就可以导致锥的细分. 所以,对 $n-1$ 维单纯形的二等分导致的锥的细分,就是锥的二等分.

设 $M = \{x \mid a \leqslant x \leqslant b\}$ 是一个 n 维的超长方体. 用一个通过点 $\frac{1}{2}(a+b)$ 的超平面垂直割该矩形的最长的边,这样把矩形 M 分成二个 n 维的超长方体. 这就是超长方体的二等分.

设 R_k 是当前的一个划分,P_k 为分枝定界过程的第 k 迭代的细分中,所选的划分元的集合. 显然,如果满足 $\beta_{k-1} = \beta(M)$ 的 $M \in R_k$ 划分是精细的,那么下界 $\beta_{k-1} \leqslant \min f(D)$ 是可以改善的. 划分是精细的定义如下：

定义 1.2.2 称划分元 $M' \subset M$ 是 M 的精细的子集,如果满足

$$\beta(M') \geqslant \beta(M), \; \alpha(M') \leqslant \alpha(M).$$

Tuy 等[63]提出了一个选择原则,即具有"界改善的"选择.

定义 1.2.3 称选择是界改善的,如果至少在迭代有限步后,P_k 满足:

$$P_k \bigcap \{M \in R_k \mid \beta(M) = \beta_{k-1}\} \neq \varnothing.$$

一般来说上界 $\alpha(M)$ 比较容易定,只要取在可行域中的最小值即可. 较精确地,可令

$$\alpha(M) = \min f(S_M), \; S_M \in M \bigcap D.$$

而下界就较麻烦. 令 M 是一个多面体,找下界 $\beta(M)$ 经常用的办法是,作 $f(x)$ 在 M 中的凸包络 φ,求 φ 在 $M \bigcap D$ 上的极小值,以它作为下界.

对于凸函数 f,其下界可以由下式给出[30,61]:

$$\beta(M) = \min f(V(P)),$$

其中 $V(P)$ 是满足 $D \bigcap M \subset P \subset M$ 的凸多面体的顶点集.

对于在单纯形上的 D.C. 函数 $f(x) = f_1(x) + f_2(x)$,其中 f_1 是凸的,f_2 是凹的. 令 φ_2 是 f_2 在 M 上的凸包络,则其下界可以由下式给出:

$$\beta(M) = \min\{f_1(x) + \varphi_2(x) \mid x \in D \bigcap M\}.$$

对于 Lipschitzian 函数 f,其下界可以由:

$$\beta(M) = \max f(S) - Ld(M),$$

其中 S 是 M 的任意一个非空,有限个点的集合,$d(M)$ 是 M 的直径,L 是 Lipschitzian 常数.

Horst 等给出了分枝定界方法求全局极值的收敛性质,其主要结论是下面定理[31]:

定理 1.2.2 设分枝定界方法中的选择是界改善的,且逐次细分

的划分元的任一下降序列 $\{M_q\}$ 满足：

$$\lim_{q \to \infty}(\alpha_q - \beta(M_q)) = 0,$$

则该方法是收敛的,即

$$\alpha := \lim_{k \to \infty}\alpha_k = \lim_{k \to \infty}f(x^k) = \min f(D) = \lim_{k \to \infty}\beta_k =: \beta.$$

§1.2.4 填充函数方法

填充函数法是由西安交通大学的葛仁溥教授等人首先提出的,
参见[16 - 20]. 以后很多学者对此方法又作了许多有益的工作和
改进.

考虑问题(1.1.2).

如果我们假设 $f(x)$ 在 R^n 上连续可微且满足强迫性条件:

假设 1 $f(x) \to +\infty$,当 $\|x\| \to +\infty$.

则问题(1.1.2)退化为

$$\min_{x \in X}f(x) \qquad\qquad (P)$$

这是因为由假设 1 知,一定存在有界闭集 X 使得 $f(x)$ 在 R^n 上的所有
极小点都在 X 的内部.

介绍几个概念.

定义 1.2.4 函数 $f(x)$ 在一极小值点 x_1^* 处的盆谷是指一连通
域 B_1^*,具有下列性质:

(i) $x_1^* \in B_1^*$;

(ii) 对于任意一点 $x \in B_1^*$ 使得 $x \neq x_1^*$ 及 $f(x) > f(x_1^*)$,存在
一条从 x 到 x_1^* 的下降路径.

若 x_1^* 是 $f(x)$ 的局部极大点,则 $-f(x)$ 在局部极小点 x_1^* 处的盆
谷称为 $f(x)$ 在局部极大点 x_1^* 的峰.

定义 1.2.5 设 x_1^* 和 x_2^* 是函数 $f(x)$ 的两个不同的极小点. 如
果 $f(x_1^*) > f(x_2^*)$,则称在 x_2^* 处的盆谷 B_2^* 比 x_1^* 处的盆谷 B_1^* 低,或

称 B_1^* 比 B_2^* 高.

一般使用填充函数方法还要求假设 2 或假设 2′.

假设 2 $f(x)$ 只有有限个极小值点.

假设 2′ $f(x)$ 只有有限个极小值.

孤立点 x_1^* 处盆谷 B_1^* 的半径定义为:

$$R = \inf_{x \notin B_1^*} \| x - x_1^* \| .$$

如果 $f(x)$ 在 x_1^* 处的 Hessian 矩阵 $\nabla^2 f(x_1^*)$ 正定,则 $R \geqslant 0$.

填充函数算法由两个阶段组成——极小化阶段和填充阶段. 这两个阶段交替使用直到找不到更好的局部极小点. 在第一阶段里,可以用经典的极小化算法如拟牛顿法和最速下降法等,寻找目标函数的一个局部极小值点 x_1^*. 然后第二阶段,主要思想是以当前极小点 x_1^* 为基础定义一个填充函数,并利用它找到 $x' \neq x_1^*$,使得

$$f(x') \leqslant f(x_1^*),$$

而后以 x' 为初始点,重复第一阶段. 重复下去一直到找不到更好的局部极小点.

设 x_1^* 是 $f(x)$ 的一个局部极小点. 文献[18]中定义:

定义 1.2.6 函数 $p(x, x_1^*)$ 称为 $f(x)$ 在局部极小点 x_1^* 处的填充函数,如果满足:

(1) x_1^* 是 $p(x, x_1^*)$ 的一个极大点,$f(x)$ 在点 x_1^* 处的盆谷 B_1^* 成为 $p(x, x_1^*)$ 的峰的一部分.

(2) $p(x, x_1^*)$ 在比 B_1^* 高的盆谷里没有稳定点.

(3) 如果存在比 B_1^* 低的盆谷 B_2^*,则存在 $x' \in B_1^*$ 使得 $p(x, x_1^*)$ 在 x' 和 x_1^* 的连线上有稳定点.

在各种文章中对填充函数的定义有所不同. 尤其(3)是一个很强的要求条件,很多文章做了改进.

葛等人定义了一个两个参数的函数

$$P(x, x_1^*, r, \rho) = \frac{1}{r + f(x)} \exp\left(-\frac{\| x - x_1^* \|^2}{\rho^2}\right).$$

后来他们注意到这个函数存在缺陷. 由于受到指数项 $\exp\left(-\frac{\| x - x_1^* \|^2}{\rho^2}\right)$ 的影响,当 ρ 太小或 $\| x - x_1^* \|$ 太大时,$P(x, x_1^*, r, \rho)$ 和 $\nabla P(x, x_1^*, r, \rho)$ 几乎都不会变化. 为了克服缺陷,葛和秦在[17]给出了七个填充函数:

$$\widetilde{P}(x, x_1^*, r, \rho) = \frac{1}{r + f(x)} \exp(-\frac{\| x - x_1^* \|}{\rho^2}),$$

$$G(x, x_1^*, r, \rho) = -\rho^2 \log[r + f(x)] - \| x - x_1^* \|^2,$$

$$\widetilde{G}(x, x_1^*, r, \rho) = -\rho^2 \log[r + f(x)] - \| x - x_1^* \|,$$

$$Q(x, x_1^*, A) = -[f(x) - f(x_1^*)] \exp(A \| x - x_1^* \|^2),$$

$$\widetilde{Q}(x, x_1^*, A) = -[f(x) - f(x_1^*)] \exp(A \| x - x_1^* \|),$$

$$\nabla E(x, x_1^*, A) = -\nabla f(x) - 2A[f(x) - f(x_1^*)](x - x_1^*),$$

$$\nabla \widetilde{E}(x, x_1^*, A) = -\nabla f(x) - A[f(x) - f(x_1^*)] \frac{x - x_1^*}{\| x - x_1^* \|}.$$

他们指出后四个函数是较好的填充函数. Liu 在文献[44](2001)中给出一个克服以上缺陷的填充函数:

$$H(x, x_1^*, a) = \frac{1}{\ln[1 + f(x) - f(x_1^*)]} - a \| x - x_1^* \|^2,$$

其中 a 充分大.

另一类关于前置点 x_0 的填充函数

$$U(x, A, h) = \eta(\| x - x_0 \|) \varphi(A[f(x) - f(x_1^*) + h])$$

是由 Ge 和 Qin (1990) 在文[20]中提出并且由 Lucidi 和 Piccialli

(2002)在文[46]中进行了深一步的讨论.

张连生,李端和 NG C. K. 对填充函数的定义进行了改进,给出了一些性质较好的函数,如文献[88]中定义的

$$p(x, x_1^*, \rho, \mu) = f(x_1^*) - \min[f(x), f(x_1^*)] - \rho \| x - x_1^* \|^2 +$$

$$\mu\{\max[0, f(x) - f(x_1^*)]\}^2.$$

该文章的另一特点是他们对填充函数的迭代搜索方法进行了详细的讨论,得到了计算方法上的一些结论. 进一步他们把改进后的填充函数用于求解非线性整数规划,建立了一个近似算法,从而为求解非线性整数规划提供了一个途径. 参见[51, 85]等.

§1.2.5   打洞函数方法

由 Levy 和 Montalco(1985)[47]首先提出打洞函数法.

考虑求解问题(P).

$$\min_{x \in X} f(x) \qquad\qquad (P)$$

文献[47]提出的打洞方法由一系列循环组成,每个循环包括两阶段: 局部极小化阶段和打洞阶段. 第一阶段极小化阶段,是由一个初始点出发,用一个求局部极小的方法,比如拟牛顿法、梯度法或共轭梯度法等,求得函数 $f(x)$ 的一个极小点 x^*. 在第二阶段打洞阶段,先定义 x^* 处的打洞函数:

$$T(x, x^*) = \frac{f(x) - f(x^*)}{[(x - x^*)^{\mathrm{T}}(x - x^*)]^\alpha}.$$

然后寻找 $T(x, x^*)$ 的零点. 即由 $T(x, x^*) = 0$ 解得 $\bar{x} \neq x^*$ 使得 $f(\bar{x}) = f(x^*)$. $\bar{x}$ 所在的盆谷比 x^* 所在的盆谷低,取 $\bar{x}$ 作为初始点开始下一个循环.

函数 $T(x, x^*)$ 的分母是强度为 α 的极锥,作用是防止原来的极小点 x^* 成为方程 $T(x, x^*) = 0$ 的根.

这样依次循环直到打洞过程找不到零点,则最后一个局部极小点即可认为是全局解. 这个方法的不足是强度参数 α 不易确定,以及非线性方程 $T(x, x^*) = 0$ 很难求解.

为了避免求解 $T(x, x^*) = 0$,Yao 在文献[81]中提出了动态打洞算法. 定义能量函数

$$E(x, x^*) = T(x, x^*) + K\int_0^{\hat{f}(x)} z\,u(z)\mathrm{d}z,$$

其中 $\hat{f}(x) = f(x) - f(x^*)$,

$$u(z) = \begin{cases} 1, z \geq 0, \\ 0, z < 0 \end{cases}$$

然后求解初始条件为 $x^* + \varepsilon$ 的动态系统:

$$\frac{\mathrm{d}x}{\mathrm{d}t} = -\frac{\partial E}{\partial x} \tag{1.2.4}$$

这里 ε 是一个扰动向量. 从给定的初始条件,能量沿着动态流递减,当系统(1.2.4)收敛到平衡状态时,即为完成打洞步骤. (1.2.4)的平衡点 $\tilde{x}$ 在比 x^* 低的盆谷里,且 $f(\tilde{x}) \leqslant f(x^*)$. 这种方法中,罚参数 K 是依赖于问题的,动态系统(1.2.4)即是非线性的又带有参数 α,求解也是比较困难的问题.

Cetin, B. C 等人根据文献[81]的思想,在[8]给出能量函数:

$$\widetilde{E}(x, x^*) = \log\left[\frac{1}{1 + \exp[-(\hat{f}(x) + a)]}\right] - \frac{3}{4}K(x - x^*)^{\frac{4}{3}}u(\hat{f}(x)).$$

和动态系统

$$\frac{\mathrm{d}x}{\mathrm{d}t} = -\frac{\partial \widetilde{E}}{\partial x}. \tag{1.2.5}$$

它的特点是系统(1.2.5)可以完成一个循环的两个步骤——极小化阶段和打洞阶段.

另外还有随机打洞算法,如[52]等.

§1.2.6 模拟退火法

模拟退火[38]算法是一种随机算法,多用于复杂的组合优化和 NP 问题. 其思想源于物理上的退火过程,数学上有"马尔可夫链"可以对它进行严格的描述. 基于马尔可夫过程理论,文献[47]理论上证明模拟退火算法以概率 1 收敛于全局最优解. 实际应用中许多参数需要调整,是一个启发式算法.

模拟退火算法基本思想如下:算法的研究对象是由一个参数集所确定的某种配置. 为了便于问题的分析,需要设计一个基于该配置的价格函数 $f(x)$,于是对配置的优化过程就转化为对 $f(x)$ 的极小化过程,$f(x)$ 的极小化过程模拟自然界的退火过程,由一个逐步冷却温度 $Temp$ 来控制,对每一个温度值,尝试一定的步骤,在每一个极小化步中,随机地选择一个新的配置并计算价格函数,这时如果 $f(x)$ 的值 f_j 小于以前的值 f_i,则选择新的配置方案;如果大于以前的值,则计算概率值: $Prob = \exp\{-\Delta f_{ij}/k * Temp\}$,这里 k 是玻尔兹曼常数,然后在 $(0,1)$ 上产生随机数 $Rand$,如果 $Rand \leqslant Prob$,则选择新的配置方案,否则仍保留原方案. 重复这些步骤,直到系统冷却不再产生更好的配置为止.

模拟退火算法的数学模型可描述为:在给定邻域结构后,模拟退火过程是从一个状态到另一个状态不断地随机游动. 此过程可以用马尔可夫链来描述. 当 t 为一确定值时,两个状态的转移概率定义为:

$$p_{ij}(t) = \begin{cases} G_{ij}(t)A_{ij}(t), & \forall j \neq i \\ 1 - \sum_{l=1, l \neq i}^{|D|} G_{ij}(t)A_{ij}(t), & j = i \end{cases} \quad (1.2.6)$$

其中 $|D|$ 表示为状态集合中状态的个数, $G_{ij}(t)$ 称为从 i 到 j 的产生概率. $G_{ij}(t)$ 表示在状态 i 时, j 状态被选取的概率. $A_{ij}(t)$ 称为接受概率, $A_{ij}(t)$ 表示状态 i 产生 j 后, 接受 j 的概率. 通常 j 被选中的概率记为:

$$G_{ij}(t) = \begin{cases} \dfrac{1}{N(x_i)}, & j \in N(i) \\ 0, & j \notin N(i) \end{cases} \quad (1.2.7)$$

而模拟退火算法中接受 j 的概率为:

$$A_{ij}(t) = \begin{cases} 1, & f(i) \geqslant f(j) \\ \exp(-\Delta f_{ij}/t), & f(i) < f(j) \end{cases} \quad (1.2.8)$$

由 (1.2.6), (1.2.7), (1.2.8) 组成模拟退火算法的主要模型.

模拟退火算法分为时齐算法和非时齐算法两类. 从理论研究可以发现, 按理论要求达到平稳点分布来应用模拟退火算法是不可能的. 时齐算法要求无穷步迭代后达到平稳分布, 而非时齐要求温度下降的迭代步为指数次的. 而从应用角度而言, 在可接受时间内达到满意解就可以了, 现有的技术尚难保证模拟退火算法得到全局最优解. 详细内容可参见文献 [47] 等.

§1.2.7 遗传算法

遗传算法是一种全局随机优化算法, 它借鉴生物界"适者生存"的自然选择思想和自然遗传机制通过选择复制和遗传因子的作用 [27], 使优化群体不断进化, 最终收敛于最优状态. 它的主要特点是:

● 群体搜索.
● 针对次数的染色体(编码)进行操作, 不需要依赖梯度信息.
● 只利用目标函数作为评优准则, 无需其他专业领域知识.
● 使用随机规则, 而不是确定规则进行搜索.

基于这些特点, 遗传算法特别适合处理那些带有多参数, 多变量, 多目标和在多区域但连通性较差的 NP - hard 问题. 并且, 在处理很多组合优化问题时, 不需要很强的技巧. 同时遗传算法与其他的启发式算法有较好的兼容性.

遗传算法包括以下主要步骤:

1) 对优化问题的解进行编码. 称一个解的编码为一个染色体,组成编码的元素称为基因.

2) 适应函数的构造和应用. 适应函数依据优化问题目标函数而定. 当适应函数确定以后,自然选择规律以适应函数值的大小决定的概率分布来确定哪些染色体适应生存,哪些被淘汰. 生存下来的染色体组成种群,生成可以繁衍下一代的群体.

3) 染色体的结合. 双亲的遗传基因结合是通过编码之间的交配达到下一代的产生.

4) 变异. 新解产生过程中可能发生基因变异,使某些解的编码发生变化,使解有更大的遍历性.

最优化问题的求解过程是从众多的解中找出最优的解. 生物进化的适者生存规律使得最具有生存能力的染色体以最大的可能生存. 所以遗传算法可以在优化问题中应用. 在优化问题中,可能点的数目是通过选择、交叉和变异的方法产生. 在选择阶段,确认某些种群产生后代,交叉操作应用于一对选择的种群来产生后代,变异则被作为后代的修正或保留的种群的修正. 像模拟退火算法一样,这类算法起源于求解组合优化问题,对连续优化问题的作用目前尚非常有限. 至今没解决的主要问题在于可能解的编码问题. 关于编码问题的讨论有这样一个观点:虽然遗传算法、模拟退火、禁忌搜索等,是具有通用性的全局最优算法,但如果不针对问题设计算法,恐怕计算时间可能非常大. 可以通过针对问题的了解,即抓住问题的特征以换取节省时间,该观点已越来越被人们接受. 具体内容可参见文献[71].

§1.3 最优性条件和收敛速度

下面给出非线性规划问题的一些预备知识.

问题(1.1.2)的局部一阶,二阶最优性条件如下.

定理 1.3.1 (一阶必要条件) 设 $f(x): D \subset R^n \to R$ 为开集 D 上的一阶连续可微函数,x^* 为 $f(x)$ 在 D 上的一个局部解,则必有 $\nabla f(x^*) = 0$.

定理 1.3.2 (二阶必要条件) 设 $f(x): D \subset R^n \to R$ 在开集 D 上二阶连续可微,x^* 为 $f(x)$ 在 D 上的一个局部解,则 $\nabla f(x^*) = 0$,$\nabla^2 f(x^*)$ 半正定.

定理 1.3.3 (二阶充分条件) 设 $f(x): D \subset R^n \to R$ 在开集 D 上二阶连续可微,x^* 为 $f(x)$ 在 D 上的一个稳定点,则当 $\nabla^2 f(x^*)$ 正定时,x^* 为 $f(x)$ 在 D 上的严格局部极小点.

对于问题(1.1.1)局部一阶,二阶最优性条件如下.

定理 1.3.4 (KKT 约束最优性的一阶必要条件) 设 $f(x)$,$g_j(x) (j \in L \cup M)$ 在包含可行域 X 的开集 $D \subset R^n$ 上一阶连续可微,x^* 是问题(1.1.1)的局部解,并且在 x^* 处有效约束梯度线性无关,则必有乘子向量 $\lambda^* \in R^p$,使得 x^*, λ^* 满足

$$\nabla f(x^*) + \sum_{j=1}^{m+p} \lambda_j^* \nabla g_j(x^*) = 0 \qquad (1.3.1)$$

$$\lambda_j^* \geqslant 0, j \in L \qquad (1.3.2)$$

$$\lambda_j^* g_j(x^*) = 0, j \in L. \qquad (1.3.3)$$

x^* 和 λ^* 称为问题(1.1.1)的 KKT 对,x^* 称为 KKT 点.

(1.3.3)式称为互补松弛条件,它表明 λ_j^* 和 $g_j(x^*)$ 不能同时不为 0. 但有可能同时为 0,即有效约束的乘子可能为 0. 如果所有有效约束的乘子均不为 0,即不存在 j 使得 $\lambda_j^* = g_j(x^*) = 0$,则说**严格互补松弛条件**成立.

严格互补松弛条件成立是一个很强且难以验证的条件. 但在很多求解约束最优问题的算法中为了保持算法的局部超线性收敛性把它作为一个假设条件. 若去掉此条件仍能保持超线性收敛速率的方

法是一项有益的工作.

若引进 Lagrange 函数

$$L(x, \lambda) = f(x) + \sum_{j=1}^{m+p} \lambda_j g_j(x), \quad \lambda_j \geqslant 0, j \in L \quad (1.3.4)$$

则(1.3.1)可表示为

$$\nabla_x L(x^*, \lambda^*) = 0.$$

定理 1.3.5 （约束最优性的二阶必要条件） 设 $f(x)$ 和 $g_j(x)$，$j \in L \bigcup M$ 在开集 $D \supset X$ 上二阶连续可微,并且约束函数全为线性函数或有效约束梯度 $\nabla g_j(x^*)(j \in L^* \bigcup M)$ 线性无关. ($L^* \subset L$ 是不等式约束在 x^* 处的有效约束指标集). x^* 和 λ^* 是问题(1.1.1)的 KKT 对,则 x^* 是问题局部解的必要条件是对于满足

$$s^T \nabla g_j(x^*) = 0 \quad (\forall j \in L^* \bigcup M)$$

的任何向量 s 都有

$$s^T \nabla_x^2 L(x^*, \lambda^*)s \geqslant 0.$$

定理 1.3.6 （约束最优性的二阶充分条件） 设 $f(x)$ 和 $g_j(x)$，$j \in L \bigcup M$ 在开集 $D \supset X$ 上二阶连续可微, x^* 和 λ^* 是问题(1.1.1)的 KKT 对. 若对满足以下条件

$$\begin{cases} s^T \nabla g_j(x^*) = 0, & j \in M \text{ 或 } j \in L^* \text{ 而 } \lambda_j^* > 0 \\ s^T \nabla g_j(x^*) \geqslant 0, & j \in L^* \text{ 而 } \lambda_j^* = 0 \end{cases}$$

的一切非零向量 s,都有

$$s^T \nabla_x^2 L(x^*, \lambda^*)s > 0$$

则 x^* 是问题(1.1.1)的一个严格局部解.

对于算法的收敛速度有以下定义.

定义 1.3.1 设由算法构造的点列 $\{x^k\}$ 收敛于 x^*. 如果存在常数 $\mu \in [0, 1)$,使得

$$\lim_{k \to \infty} \frac{\| x^{k+1} - x^* \|}{\| x^k - x^* \|} \leqslant \mu$$

则称 $\{x^k\}$ 的收敛速率至少是线性的简称线性的；如果

$$\lim_{k \to \infty} \frac{\| x^{k+1} - x^* \|}{\| x^k - x^* \|} = 0$$

则称 $\{x^k\}$ 是超线性收敛的.

第二章 打洞函数和填充函数的统一

本章提出了两类变换函数,它们既具有打洞函数的特点又具有填充函数的性质,因此用于全局最优化的填充函数法和打洞函数法在一定条件下是可以统一的. 特别是打洞函数具有填充函数的主要性质,而很多填充函数可以认为具有打洞函数的特点.

§2.1 引言

很多自然科学,经济和工程学上的最新发展与进步都需要求解下面 n 维多极值点的数学规划问题:

$$\min \{ f(x) : x \in \Omega \} \qquad (2.1.1)$$

其中 $f : R^n \to R$, $\Omega \subseteq R^n$ 是一个充分大的闭区域. 因此全局优化问题的研究成为一个被高度关注的热点问题之一. 但是由于很可能在一个全局优化问题里有许多局部极小点,因此全局优化问题不能简单地用通常意义上的求解目标函数局部极小的方法. 这些多极小值点的函数一般会导致两个比较困难的问题:第一个是怎样离开一个极小值点到一个更优的极小点;第二个是怎样判断当前的极小值点是否为全局最优解.

由 Levy 和 Montalco(1985)[47]提出的打洞函数法,和由 Ge 和 Qin[17]介绍的填充函数的算法是两个求解全局优化问题的实用方法.

打洞算法由两个阶段组成——极小化阶段和打洞阶段. 这两个阶段交替使用直到找不到更小的局部极小点. 在第一阶段中,可以用经典的极小化算法如拟牛顿法、共轭梯度法或最速下降法,寻找目标函数的局部极小值点. 在第二阶段中,主要思想是以当前极小点为基础定义一个辅助函数——打洞函数并寻找它异于 x^* 的零点. 在 Levy 和 Montalvo(1985)[47]的论文里,定义打洞函数为:

$$T(x, x^*) = \frac{f(x) - f(x^*)}{[(x - x^*)^\mathrm{T}(x - x^*)]^\alpha} \qquad (2.1.2)$$

如果可以找到 $\bar{x} \neq x^*$ 使得 $T(\bar{x}, x^*) = 0$,它等价于 $f(\bar{x}) = f(x^*)$,那么 $\bar{x}$ 可作为下一个极小化阶段的初始点. (2.1.2)式的分母是点 x^* 处强度为 α 的极点,其作用是防止 x^* 成为打洞函数的零点.

填充函数法类似于打洞算法. 区别在于第二阶段的辅助函数不同,它用填充函数寻找点 $\bar{x}$. 就表达式来看有各种各样的填充函数. 下面的函数是 Han etc. (2001)在文章[25]中所应用的填充函数:

$$P(x, r, \rho) = \frac{1}{r + f(x)} \exp\left(-\frac{\|x - x^*\|^2}{\rho}\right) \qquad (2.1.3)$$

另一类关于前置点 x_0 的填充函数

$$U(x, A, h) = \eta(\|x - x_0\|)\varphi(A[f(x) - f(x^*) + h]) \qquad (2.1.4)$$

是由 Ge 和 Qin (1990)在文[20]中提出并且由 Lucidi 和 Piccialli (2002)在文[46]中进行了进一步的讨论. 他们都假设 $f(x)$ 只有有限个极小值点,本文将此假设弱化为 $f(x)$ 只有有限个不同的极小值.

无论填充函数的表达式如何,如(2.1.3)和(2.1.4)式,在一定的假设下具有以下共同的性质:

(i) x^* 是填充函数的极大点.

(ii) 在区域 $\{x \mid f(x) \geqslant f(x^*)\}$ 上,填充函数沿射线 $x^* \to x$ 没有极小点.

(iii) 如果 $f(x^*)$ 不是全局最优值,填充函数在连线 $x^* \to x'$ 上有一个极小值点. 这里 x' 所在的盆谷比 x^* 所在的盆谷更低.

本文的目的是求问题(2.1.1)的一个全局最优解. 对目标函数 $f(x)$ 有以下假设:

A1 $f(x)$ 在 R^n 上连续可微, 并且存在常数 $K > 0$ 使得 $\| \nabla f(x) \| \leqslant K, \ \forall x \in R^n$;

A2   当 $\| x \| \to +\infty$ 时, $f(x) \to +\infty$, 即 $f(x)$ 是一个强制函数;

A3 $f(x)$ 只有有限个不同的极小值.

根据假设 A2 可知, 目标函数的所有极小点都在一个有界闭区域 Ω 内. 所以对于 n 维无约束全局最优化问题 $\min \{ f(x) : x \in R^n \}$, 我们只需考虑问题(2.1.1). 这样我们记

$$\overline{D} = \max_{x \in \Omega} \| x - x^* \| \qquad (2.1.5)$$

本文主要涉及和讨论我们提出的两类变换函数. 我们将证明它们既具有打洞函数的特点, 又具有填充函数的性质, 这样打洞函数也可以被看做填充函数. 当然, 很多填充函数能被看做打洞函数, 如表达式 (2.1.4). 因此打洞函数和填充函数在某些场合下可以统一. 事实上, 用填充函数或者打洞函数找到第一阶段的一个新的初始点的计算过程可以是非常相似的. 基于以上理由, 这篇论文里提到的变换函数被统称为 T-F 函数.

本章是这样安排的: 在第 2 节, 我们讨论第一类 T-F 函数的定义和性质, 这类函数的分母比(2.1.2)式的要简单而分子是一些满足一定条件的抽象函数. 然后分母是另一种抽象函数的第 2 类 T-F 函数将在第 3 节提出. 在最后一部分我们将报告一些计算的例子, 这些算例说明算法是有效的.

§2.2   第一类 T-F 函数及其性质

本节我们介绍一族变换函数.

首先我们有几个约定：假设已找到问题 (2.1.1)的一个局部极小点 x^*；参数 $\tau \geqslant 1$；又假设 h 满足：

$$0 < h < f(x^*) - f(x_G^*) \qquad (2.2.1)$$

这里 x_G^* 是 $f(x)$ 的一个全局最优点.

根据 (2.2.1)式可知 h 在一定程度上是一个依赖于问题的参数. 本节中设抽象函数 $\phi(t)$ 具有以下性质：

(i) $\phi(0) = 0$；

(ii) $\forall t \in [-t_1, \infty)$, $\phi'(t) > 0$ (这里 $t_1 \geqslant 0$)；

(iii) $\lim\limits_{t \to +\infty} \dfrac{t\phi'(t)}{\phi(t)} = 0$.

有许多函数拥有上述性质,例如

- $\phi(t) = \ln(1+t)$
- $\phi(t) = t/(1+t)$
- $\phi(t) = \arcsin(t/(1+t))$
- $\phi(t) = 1 - e^{-t}$

第一类变换函数的定义如下：

$$T_1(x, x^*, \tau) = \frac{\phi(\tau[f(x) - f(x^*) + h])}{\| x - x^* \|} \qquad (2.2.2)$$

其中 τ 和 h 是参数,其取法如约定. 显然,如果有 $\bar{x} \neq x^*$ 使得 $T_1(\bar{x}, x^*, \tau) = 0$,则 $f(\bar{x}) < f(x)$. 几何上,点 $(\bar{x}, f(\bar{x}))$ 所在的盆谷低于点 $(x^*, f(x^*))$ 所在的盆谷. 所以函数 $T_1(\bar{x}, x^*, \tau)$ 具有打洞函数的特点,因而是一族打洞函数. 由于它比文献[47]提出的打洞函数在形式上不同,我们也可以称它为变形打洞函数.

为简洁起见,用 $\hat{f}$ 表示 $f(x) - f(x^*) + h$, $\hat{f}_1$ 表示 $f(x_1) - f(x^*) + h$. 等等.

参数 h 在某种意义上也可以认为是人为控制的容许值,这是由于在计算过程中我们可以调节 h 的取值. 如果虽然 $h>0$ 但已充分小仍然不能找到函数 $T_1(x, x^*, \tau)$ 的零点,则可认为当前的局部极小点是全局最优点. 从这个意义上参数 h 的作用可以部分地解决文章开头提出的全局最优化问题所带来的第二个困难.

函数 $T_1(x, x^*, \tau)$ 关于 x 的梯度是:

$$\nabla T_1(x, x^*, \tau) = \frac{1}{\|x-x^*\|} \{\tau \phi'(\tau \hat{f}) \nabla f(x) -$$

$$\frac{\phi(\tau \hat{f})}{\|x-x^*\|^2}(x-x^*) \} \qquad (2.2.3)$$

显然, $\lim\limits_{\|x-x^*\| \to 0} T_1(x, x^*, \tau) = +\infty$. 因此可以广义地认为 x^* 是变换函数 $T_1(x, x^*, \tau)$ 的极大值点.

如果用 $N(x^*)$ 代表点 x^* 的邻域,则由于 $f(x)$ 和 $\phi(t)$ 的可微性知函数 $T_1(x, x^*, \tau)$ 在区域 $\Omega \setminus N(x^*)$ 上是可微的.

定理 2.2.1 如果点 x^* 是 $f(x)$ 的一个局部极小点并且函数 $\phi(t)$ 满足性质(i)—(iii),则对充分大的 τ 函数 $T_1(x, x^*, \tau)$ 在区域 $S_1 = \{x \mid f(x) \geqslant f(x^*), x \in \Omega \setminus x^*\}$ 上没有稳定点.

证明 任取 $x \in S_1$,即 $f(x) \geqslant f(x^*)$ 并且 $x \neq x^*$. 现在我们证明

$$(x-x^*)^T \nabla T_1(x, x^*, \tau) < 0, \forall x \in S_1 \qquad (2.2.4)$$

根据 $T_1(x, x^*, \tau)$ 的梯度表达式,即(2.2.3)式可得

$$(x-x^*)^T \nabla T_1(x, x^*, \tau) = \frac{1}{\|x-x^*\|} \{\tau \phi'(\tau \hat{f}) \nabla f(x)^T(x-$$

$$x^*) - \phi(\tau \hat{f})\}.$$

注意到

$$(x-x^*)^T \nabla f(x) \leqslant \|x-x^*\| \|\nabla f(x)\| \leqslant M\overline{D},$$

由 $\phi(t)$ 的性质（ii）和假设 A1，我们可得下面的不等式：

$$(x - x^*)^{\mathrm{T}} \nabla T_1(x, x^*, \tau)$$

$$\leqslant \frac{1}{\| x - x^* \|} \{\tau \phi'(\tau \hat{f}) M \overline{D} - \phi(\tau \hat{f})\}.$$

如果想要 $(x - x^*)^{\mathrm{T}} \nabla T_1(x, x^*, \tau) < 0$ 成立，只需 τ 充分大使得

$$\frac{\tau \phi'(\tau \hat{f})}{\phi(\tau \hat{f})} \leqslant \frac{1}{M \overline{D}}.$$

根据 $\phi(t)$ 的性质（iii）上式显然成立. 从而（2.2.4）式成立.

既然（2.2.4）式成立，作为结果必有 $\forall x \in S_1, \nabla T_1(x, x^*, \tau) \neq 0$.

不等式（2.2.4）也说明下面的结论是正确的.

推论 2.2.1 如果 $x \in S_1$，即 $f(x) \geqslant f(x^*)$，则 $x - x^*$ 是 $T_1(x, x^*, \tau)$ 在点 x 处的严格下降方向.

定理 2.2.1 和推论 2.2.1 揭示了 $T_1(x, x^*, \tau)$ 的填充性质. 下面的定理进一步说明了函数 $T_1(x, x^*, \tau)$ 的其他基本的填充性质.

定理 2.2.2 若 x^* 不是 $f(x)$ 的全局最优解并且函数 $\phi(t)$ 具有性质（i）—（iii），则函数 $T_1(x, x^*, \tau)$ 在区域 $S_2 = \{x \mid f(x) < f(x^*), x \in \Omega\}$ 上有一个极小点 $\tilde{x}$.

证明 由于 x^* 不是 $f(x)$ 的全局最优解，所以 S_2 非空. 注意到（2.2.1）式和 $\phi(t)$ 的性质，我们有 $\phi(\tau[f(x_G^*) - f(x^*) + h]) < 0$. 因此，根据 $T_1(x, x^*, \tau)$ 的定义有 $T_1(x_G^*, x^*, \tau) < 0$ 成立. 另外，由于连续性，函数 $T_1(x, x^*, \tau)$ 在有界闭区域 $\Omega \backslash N(x^*)$ 上有一个全局极小值点 $\tilde{x}$. 从而有

$$T_1(\tilde{x}, x^*, \tau) \leqslant T_1(x_G^*, x^*, \tau) < 0 \qquad (2.2.5)$$

由 $T_1(x, x^*, \tau)$ 的定义知，不等式（2.2.5）可得 $\phi(\tau[f(\tilde{x}) - f(x^*) + h]) < 0$，又由 $\phi(t)$ 的性质知 $f(\tilde{x}) < f(x^*)$ 成立，即 $\tilde{x} \in S_2$.

显然有 $\lim\limits_{x \to x^*} T_1(x, x^*, \tau) = +\infty$. 若用 $\partial\Omega$ 代表 Ω 的边界, 则根据假设 A2, 对 $x \in \partial\Omega$ 有 $f(x) > f(x^*)$ 成立, 这样在边界 $\partial\Omega$ 上 $T_1(x, x^*, \tau) > 0$. 这说明函数 $T_1(x, x^*, \tau)$ 在 $N(x^*)$ 和 Ω 的边界上没有极小点. 因此, $\tilde{x}$, $T_1(x, x^*, \tau)$ 的极小点是区域 $\Omega \backslash N(x^*)$ 的内点, 而且函数 $T_1(x, x^*, \tau)$ 在点 $\tilde{x}$ 是可微的. 所以有下列推论.

推论 2.2.2 如果 x^* 不是问题的全局最优解, 则 $T_1(x, x^*, \tau)$ 在区域 S_2 上至少有一个稳定点.

下面的定理说明了 $T_1(x, x^*, \tau)$ 的一个特性, 距离 x^* 越远, 函数值越小.

定理 2.2.3 假设 $\varepsilon > 0$, 点 $x_1, x_2 \in S_1$ 并且 $\| x_1 - x^* \| \geqslant \| x_2 - x^* \| + \varepsilon$.

(a) 如果 $\lim\limits_{t \to +\infty} \phi(t) = B$ (B 是一个常数), 那么对充分大的 τ 有

$$T_1(x_1, x^*, \tau) < T_1(x_2, x^*, \tau) \qquad (2.2.6)$$

(b) 如果

$$\lim\limits_{t \to +\infty} \frac{\phi(t)}{\ln (1 + t)} = C$$

(这里 C 是一个正常数), 则对充分大的 τ 亦有不等式 (2.2.6) 式成立.

证明 设 $x_1, x_2 \in S_1$, 即 $f(x_1) \geqslant f(x^*)$, $f(x_2) \geqslant f(x^*)$.

(a) 由假设 $\lim\limits_{t \to +\infty} \phi(t) = B$, 所以

$$\lim\limits_{\tau \to +\infty} \frac{\phi(\tau \hat{f}_2)}{\phi(\tau \hat{f}_1)} = 1.$$

$\| x_1 - x^* \| \geqslant \| x_2 - x^* \| + \varepsilon$ 我们可得

$$\frac{\| x_2 - x^* \|}{\| x_1 - x^* \|} \leqslant 1 - \frac{\varepsilon}{\| x_1 - x^* \|} \leqslant 1 - \frac{\varepsilon}{D}.$$

因此, 当 τ 足够大后有

$$\frac{\phi(\tau \hat{f}_2)}{\phi(\tau \hat{f}_1)} > \frac{\| x_2 - x^* \|}{\| x_1 - x^* \|} \tag{2.2.7}$$

即 (2.2.6)式成立.

(b) 如果 $\phi(t) = \ln(1+t)$. 因为当常数 $C_1 > 0, C_2 > 0,$

$$\lim_{x \to +\infty} \frac{\ln(1+C_1 x)}{\ln(1+C_2 x)} = \lim_{x \to +\infty} \frac{C_1(1+C_2 x)}{C_2(1+C_1 x)} = 1,$$

所以

$$\lim_{\tau \to +\infty} \frac{\ln(1+\tau \hat{f}_2)}{\ln(1+\tau \hat{f}_1)} = 1.$$

类似于(2.2.7)式的证明,对充分大的 τ

$$\frac{\ln(1+\tau \hat{f}_2)}{\ln(1+\tau \hat{f}_1)} > \frac{\| x_2 - x^* \|}{\| x_1 - x^* \|}.$$

即 $T_1(x_1, x^*, \tau) < T_1(x_2, x^*, \tau)$.

如果 $\phi(t) \neq \ln(1+t)$,但是

$$\lim_{t \to +\infty} \frac{\phi(t)}{\ln(1+t)} = C,$$

那么

$$\lim_{\tau \to +\infty} \frac{\phi(\tau \hat{f}_2)}{\phi(\tau \hat{f}_1)} = \lim_{\tau \to +\infty} \left[\frac{\phi(\tau \hat{f}_2)}{\ln(1+\tau \hat{f}_2)} \cdot \frac{\ln(1+\tau \hat{f}_1)}{\phi(\tau \hat{f}_1)} \cdot \right.$$

$$\left. \frac{\ln(1+\tau \hat{f}_2)}{\ln(1+\tau \hat{f}_1)} \right] = 1.$$

因此,当 τ 充分大时不等式 (2.2.7)因而(2.2.6)式成立.

现在给出一些解释. 定理 2.2.1 和定理 2.2.2 说明如果 x^* 不是问题 (2.1.1) 的全局解,则当参数 τ 和 h 取适当的值后,函数 $T_1(x,$

x^*, τ) 可以确定点 $\tilde{x}$ 使得 $f(\tilde{x}) < f(x^*)$. 而且根据定理 2.2.3,点 x 离开 x^* 越远,函数 $T_1(x, x^*, \tau)$ 的值越小. 所以在第二个阶段我们没有必要找出 $T_1(x, x^*, \tau)$ 的极小值点 $\tilde{x}$,只要找到 x_1 使得 $f(x_1) < f(x^*)$ 就足够了. 从几何上看,这时 $T_1(x, x^*, \tau)$ 已经使得目标函数 $f(x)$ 离开了点 $(x^*, f(x^*))$ 所在的盆谷到达了更低的,点 $(\tilde{x}, f(\tilde{x}))$ 所在的盆谷. 从而 x_1 即为下一个极小化阶段的初始点. 辅助函数的特性非常有助于于求解全局最优化问题. 然而,对参数 τ 的取值掌握的是否合适也是涉及辅助函数有效性的重要问题. 一般地讲,τ 应该适当的大而 h 应该逐步减小,直到 $h < \varepsilon$,这里 $\varepsilon > 0$ 是一个容许度. 如果 h 充分小,小于 ε 并且 $T_1(x, x^*, \tau)$ 在 S_2 上仍没有找到极小点,则可认为最后一个局部极小点为全局解.

§2.3　第二类 T-F 函数及其性质

本节给出另一类变换函数. 即:

$$T_2(x, x^*, \alpha) = \frac{f(x) - f(x^*) + h}{[\psi(\|x - x^*\|)]^\alpha} \qquad (2.3.1)$$

其中 $\alpha > 0$ 是一个依赖于问题的参数,h 依然满足 (2.2.1) 式.

本节假设函数 $\psi(t)$ 有下列性质:

（Ⅰ）$\psi(0) = 0$;

（Ⅱ）对 $t \geq 0$, $\psi'(t) > 0$,并且 $\psi'(t)$ 单调不增.

几个满足上述性质的函数如下:

● $\psi(t) = t$

● $\psi(t) = \arctan t$

● $\psi(t) = \tanh t$

由 $\psi(t)$ 的性质和(2.3.1)式知,函数 $T_2(x, x^*, \alpha)$ 具有打洞函数的特性,即若有 $\bar{x} \neq x^*$ 使得 $T_2(\bar{x}, x^*, \alpha) = 0$,则 $f(\bar{x}) < f(x)$.

正如 $T_1(x, x^*, \tau)$ 那样,函数 $T_2(x, x^*, \alpha)$ 也具有填充函数的某

些性质.

首先,由函数 $\psi(t)$ 的性质知,$\lim\limits_{x \to x^*} T_2(x, x^*, \alpha) = +\infty$,因而广义地看 x^* 是 $T_2(x, x^*, \alpha)$ 的极大值点. 下面的几个定理进一步证明了 $T_2(x, x^*, \alpha)$ 的填充性质.

定理 2.3.1 设 x^* 是 $f(x)$ 的一个极小点并且函数 $\psi(t)$ 满足性质（Ⅰ）—（Ⅱ）. 如果

$$\alpha > \frac{K}{h} \frac{\psi(\overline{D})}{\psi'(\overline{D})} \tag{2.3.2}$$

则 $T_2(x, x^*, \alpha)$ 在区域 S_1 上没有稳定点.

证明 由(2.3.1)式,函数 $T_2(x, x^*, \alpha)$ 关于 x 的梯度向量为:

$$\nabla T_2(x, x^*, \alpha) = \frac{1}{[\psi(\|x-x^*\|)]^\alpha}[\nabla f(x) - \alpha(f(x) - f(x^*) +$$

$$h) \times \frac{\psi'(\|x-x^*\|)}{\psi(\|x-x^*\|)} \frac{x-x^*}{\|x-x^*\|}] \tag{2.3.3}$$

任取 $x \in S_1$,即 $f(x) \geqslant f(x^*)$ 且 $x \neq x^*$. 另外为简单起见,用 $\psi(\cdot)$ 表示函数 $\psi(\|x-x^*\|)$.

为了证明 $\nabla T_2(x, x^*, \alpha) \neq 0$,只需证明

$$(x-x^*)^T \nabla T_2(x, x^*, \alpha) = \frac{1}{[\psi(\cdot)]^\alpha}[(x-x^*)^T \nabla f(x) -$$

$$\alpha \|x-x^*\| (f(x) - f(x^*) + h) \frac{\psi'(\cdot)}{\psi(\cdot)}] \neq 0.$$

注意到 $\|\nabla f(x)\| \leqslant M$ 和 $f(x) - f(x^*) \geqslant 0$,并且设 θ 代表向量 $(x-x^*)$ 和 $\nabla f(x)$ 之间的夹角. 那么有

$$(x-x^*)^T \nabla T_2(x, x^*, \alpha)$$

$$\leqslant \frac{1}{[\psi(\cdot)]^\alpha}[K \|x-x^*\| \cos\theta - \alpha \|x-x^*\| h \frac{\psi'(\cdot)}{\psi(\cdot)}].$$

如果不等式（2.3.2）成立，基于 $\psi(t)$ 的性质（Ⅰ）和（Ⅱ），我们有

$$(x-x^*)^{\mathrm{T}} \nabla T_2(x, x^*, \alpha) <$$

$$\frac{M\|x-x^*\|}{[\psi(\|x-x^*\|)]^\alpha}\left[\cos\theta - \frac{\psi(\overline{D})}{\psi(\|x-x^*\|)}\frac{\psi'(\|x-x^*\|)}{\psi'(\overline{D})}\right] \leqslant 0.$$

证毕.

定理 2.3.2 若 x^* 不是问题（2.1.1）的全局最优解，并且 $\psi(t)$ 满足性质（Ⅰ）—（Ⅱ），则 $T_2(x, x^*, \alpha)$ 在区域 S_2 上有一个极小点，此点同时也是一个稳定点.

证明 因为 x^* 不是全局极小点，所以根据预设条件（2.2.1）式，有全局极小点 x_G^* 使得 $f(x_G^*)-f(x^*)+h<0$. 而且当 $\forall x \neq x^*$ 时 $\psi(\|x-x^*\|)>0$，因而 $T_2(x_G^*, x^*, \alpha)<0$.

根据 $T_2(x, x^*, \alpha)$ 的连续性，知其在有界闭域 $\Omega/N(x^*)$ 上有最小点 $\hat{x}$，并且由假设 A2 和极限 $\lim\limits_{\|x-x^*\|\to 0} T_2(x, x^*, \alpha) = +\infty$ 知，点 $\hat{x}$ 位于区域的内部. 因此 $T_2(x, x^*, \alpha)$ 在点 $\hat{x}$ 处是可微的. 这就是说，$\hat{x}$ 是函数 $T_2(x, x^*, \alpha)$ 的稳定点.

现在证明 $\hat{x}$ 落在区域 S_2 上.

易知

$$T_2(\hat{x}, x^*, \alpha) \leqslant T_2(x_G^*, x^*, \alpha) < 0$$

这说明 $f(\hat{x})-f(x^*)+h<0$. 所以 $f(\hat{x})<f(x^*)$，即 $\hat{x}\in S_2$.

定理 2.3.3 如果 $x_1, x_2 \in S_1$ 而且有 $\varepsilon>0$ 使得 $\|x_1-x^*\| \geqslant \|x_2-x^*\|+\varepsilon$，则对适当大的 α 有

$$T_2(x_1, x^*, \alpha) < T_2(x_2, x^*, \alpha).$$

证明 因为 $\psi'(t)>0$，所以有 $\delta=O(\varepsilon)>0$ 使得

$$\psi(\|x_1-x^*\|) \geqslant \psi(\|x_2-x^*\|)+O(\varepsilon).$$

进而有

$$\frac{\psi(\|x_2 - x^*\|)}{\psi(\|x_1 - x^*\|)} \leqslant 1 - \frac{O(\varepsilon)}{\psi(\overline{D})}.$$

另一方面，由假设 A1 和 $x_2 \in S_1$ 有

$$\frac{f(x_2) - f(x^*) + h}{f(x_1) - f(x^*) + h} \geqslant \frac{h}{K\overline{D} + h}.$$

所以当

$$\alpha > \frac{\ln h - \ln(K\overline{D} + h)}{\ln[\psi(\overline{D}) - O(\varepsilon)] - \ln[\psi(\overline{D})]}$$

时，

$$\frac{f(x_2) - f(x^*) + h}{f(x_1) - f(x^*) + h} \geqslant \frac{h}{K\overline{D} + h} > \left(1 - \frac{O(\varepsilon)}{\psi(\overline{D})}\right)^{\alpha}$$

$$\geqslant \left[\frac{\psi[\|x_2 - x^*\|]}{\psi[\|x_1 - x^*\|]}\right]^{\alpha},$$

即 $T_2(x_1, x^*, \alpha) < T_2(x_2, x^*, \alpha)$.

§2.4 数值结果

前面的内容讨论了两类求解全局问题的辅助函数以及它们的性质，得出了一些新的结论和观点. 虽然本章主要侧重于理论分析而非计算，但是我们仍做了一些数值试验以说明在全局最优化求解过程的第二阶段可将打洞函数作为填充函数应用. 所用的辅助函数是

$$\frac{\ln(1 + \tau(f(x) - f(x^*) + h))}{\|x - x^*\|}.$$

下面为理论算法：

1. 初始化

选取 $\varepsilon > 0$ 和 $h > 0$ 作为算法终止的容许参数.

取 $\tau > 0$ and $M > 0$.

选取方向 e_i, $i = 1, 2, \cdots, k_0$，这里 $k_0 > 2n$，n 是变量的维数.

取初始点 $x_1^0 \in \Omega$.

2. 主迭代步

1° 以 x^0 为初始点，用下降算法求出一个原问题的局部极小点. 令 x_k^* 为当前的局部极小点. 令 $i = 1$.

2° 如果 $i > k_0$ 并且 $h < \varepsilon$ 则停，x_k^* 是一个全局极小点.

3° 如果 $i > k_0$ 并且 $h > \varepsilon$ 那么，令 $r = r/10$ 和 $i = 1$ 否则转 4°.

4° 令 $\bar{x}_k^* = x_k^* + \delta e_i$（这里，$\delta$ 是一个非常小的正数），如果 $f(\bar{x}_k^*) < f(x_k^*)$ 则令 $k = k+1$，$x_k^0 = \bar{x}_k^*$ 并转到 1°；否则转 5°.

5° 令

$$T(y, x_k^*, \tau, h) = \frac{\ln(1 + \tau(f(y) - f(x_k^*) + h))}{\|y - x^*\|}$$

及 $y_0 = \bar{x}_k^*$. 转内循环.

3. 内循环

1° $y_{m+1} = \varphi(y_m)$，这里 φ 为迭代函数. 它以 y_0 为初始点，对函数 $T(y, x_k^*, \tau, h)$ 施行下降局部算法，如拟牛顿法，梯度法等.

2° 如果 $\|y_{m+1} - x_1^0\| \geq M$，则置 $i = i+1$，转到主迭代步 2°.

3° 如果 $f(y_{m+1}) \leq f(x_k^*)$ 则令 $k = k+1$，$x_k^0 = y_{m+1}$ 并转主迭代步 1°，否则令 $m = m+1$ 转 1°.

从算法的理论描述上我们可以看到，我们并不是一定要找到变换函数的零点，而是在使用局部迭代算法最小化变换函数时，一旦发现函数值小于当前最小值的点时我们就终止最小化变换函数的迭代，转为以这个点为初始点，最小化目标函数的迭代，内循环就完成

这个任务. 进入主循环后, 我们使用局部的迭代算法最小化目标函数, 得到更小的函数值, 然后在主迭代的第 5 步, 构造新的变换函数, 转到内循环. 在最小化变换函数迭代的初始点的选择上, 我们选择在当前最小点附近逐次按 k_0 各方向轻微扰动, 也就是这些点: $x_k^* + \delta e_i$, $i = 1, 2, \cdots, k_0$, 当以这些点为初始点, 最小化变换函数的迭代失败时, 我们判断变换函数中的参数 h 是否足够小, 如果小于一定的值我们终止整个算法, 认为不存在更小的局部最小点了, 当前的最小点已经是全局最小点, 否则缩小参数 h, 算法重新开始, 直到 h 一定的小, 并且从 k_0 个出发最小化变换函数都失败, 则算法终止. 在扰动方向的选择上我们通常选取 $2n$ 个坐标方向. 算法的总体结构是这样的, 在选择参数时, 我们让 ε 充分的小, h 尽可能的大一点, 这样既不会漏掉全局最小点同时也会让算法尽快地找到全局最小点, 数值试验也证明了这种策略的可行性, 如果想让算法更简洁, 只要在初始化阶段选择参数 h 时, 让它充分的小就可以了, 这样当 i 累计超过 k_0 时, 算法终止. 内循环的第 2 步中 $M > 0$ 是一个控制量, $\| y_{m+1} - x_1^0 \| \geqslant M$ 表示搜索到边界, (实际计算时经常出现这种情况) 这时要么已经得到最优解, 要么要重新找搜索方向. 在估计算法的计算量时, 它严重地依赖我们所选择的局部最小化算法的计算量, 因为每一次迭代可分为两个局部最小化阶段: 最小化变换函数和最小化目标函数, 因此可以认为每一步迭代的计算量大体上是局部最小化计算量的两倍. 如果算法迭代的 k 步终止他的计算量大体上是 $2k$ 倍的局部最小化算法的计算量, 所以我们在选择局部最小化算法时, 那些计算量小的算法是我们的首选, 我们常常选择最速下降法.

下面的测试函数都是常用的函数, 因为它们经常用于验证求解全局最优问题的各种方法的有效性. 这些函数的计算用 Fortran90 编程, 从计算量的角度考虑, 我们使用最速下降法, 来解决局部最小化问题, 参数的选取上, 我们让 $h = 1$, $\varepsilon = 0.00001$, 让 e_i 是 $2n$ 个坐标方向, 最后我们成功地应用了本文提出的理论方法寻找到了下面各个问题的全局极小值.

测试函数

(i) 6-hump camel back 函数：

$$f(x) = 4x_1^2 - 2.1x_1^4 + \frac{1}{3}x_1^6 - x_1x_2 - 4x_2^2 + 4x_2^4,$$

$$-3 \leqslant x_1, x_2 \leqslant 3$$

其全局最优解为：$x_G^* = (0.089\,8, 0.712\,7)$ 或 $(-0.089\,8, -0.712\,7)$ 并且 $f_G^* = -1.031\,6$. 我们只求出了 3 个局部最小点就达到了全局最优,其计算量大体是 6 倍的最速下降法的平均计算量. 见表 1.

(ii) Goldstein and Price 函数 [20]：

$$f(x) = [1 + (x_1 + x_2 + 1)^2(19 - 14x_1 + 3x_1^2 - 14x_2 +$$

$$6x_1x_2 + 3x_2^2)] \times [30 + (2x_1 - 3x_2)^2(18 -$$

$$32x_1 + 12x_1^2 + 48x_2 - 36x_1x_2 + 27x_2^2)],$$

$$-3 \leqslant x_1, x_2 \leqslant 3$$

其全局最优解为：$x_G^* = (0.000\,0, -1.000\,0)$；最优值：$f_G^* = 3.000\,0$. 我们只求出了 2 个局部最小点就达到了全局最优,其计算量大体是 4 倍的最速下降法的平均计算量. 见表 2.

(iii) Treccani 函数[20]：

$$f(x) = x_1^4 + 4x_1^3 + 4x_1^2 + x_2^2$$

$$-3 \leqslant x_1, x_2 \leqslant 3$$

其全局最优解为：$x_G^* = (0.000\,0, 0.000\,0)$ 或 $(-2.000\,0, 0.000\,0)$；最优值：$f_G^* = 0.000\,0$. 我们也是出了 3 个局部最小点就达到了全局最优,其计算量大体是 6 倍的最速下降法的平均计算量. 见表 3.

(iv) Rastrigin[44]：

$$f(x) = x_1^2 + x_2^2 - \cos(18x_1) - \cos(18x_2)$$

$$-1 \leqslant x_1,\ x_2 \leqslant 1$$

其全局最优解为：$x_G^* = (0.000\ 0,\ 0.000\ 0)$；最优值：$f_G^* = -2.000\ 0$. 我们求出了 4 个局部最小点就达到了全局最优，其计算量大体是 8 倍的最速下降法的平均计算量. 见表 4.

（v）2-dimensional 函数[32]：

$$f(x) = [1 - 2x_2 + c\sin(4\pi x_2) - x_1]^2 + [x_2 - 0.5\sin(2\pi x_1)]^2$$

$$-10 \leqslant x_1,\ x_2 \leqslant 10$$

这里 $c = 0.2,\ 0.5,\ 0.05$.

其全局最优值为：对所有 c，$f_G^* = 0.000\ 0$. 我们只求出了 4 个局部最小点就达到了全局最优，其计算量大体是 8 倍的最速下降法的平均计算量. 见表 5.

（vi）2-dimensional Shubert 函数[20]：

$$f(x) = \left\{ \sum_{i=1}^{5} i\cos[(i+1)x_1 + i] \right\}$$

$$\left\{ \sum_{i=1}^{5} i\cos[(i+1)x_2 + i] \right\}$$

$$-10 \leqslant x_1,\ x_2 \leqslant 10$$

此问题有 700 多个局部极小点，其一个全局最优解为：$x_G^* = (-1.425\ 2,\ -0.800\ 3)$；最优值：$f_G^* = -186.730\ 9$. 我们只求出了 2 个局部最小点就达到了全局最优，其计算量大体是 4 倍的最速下降法的平均计算量. 见表 6.

（vii）n-dimensional Sine-square 函数[20]：

$$f(x) = \frac{\pi}{n} \left\{ 10\sin^2(\pi x_1) + \sum_{i=1}^{n-1} [(x_i - 1)^2 \right.$$

$$\left. (1 + 10\sin^2(\pi x_{i+1})] + (x_n - 1)^2 \right\}$$

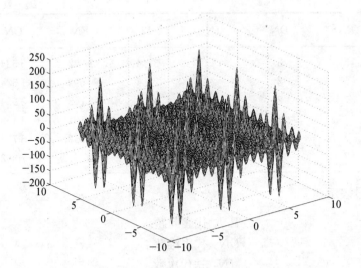

$$-10 \leqslant x_i \leqslant 10 \quad i = 1, 2, \cdots, n$$

对 $n = 2, 3, 6, 10$ 分别作了计算. 它们的最优解都是: $x_G^* = (1.000\ 0, 1.000\ 0, \cdots, 1.000\ 0)$, 最优值: $f_G^* = 0.000\ 0$.

我们只求出了 3 个局部最小点就达到了全局最优, 其计算量大体是 6 倍的最速下降法的平均计算量. 见表 7.

计算结果见如下结果表:

结果表

PN	DN	IN	TI	FN	GN
1	2	3	13.3	1 714	3 654
2	2	2	12.11	446	1 123
3	2	3	36.22	3 996	7 843
4	2	4	45.65	2 264	4 876

续　表

PN	DN	IN	TI	FN	GN
5	2	4	37.4	2 302	6 534
6	2	2	27.5	692	1 534
7	2	3	65.8	2 768	3 934
7	4	3	50	2 101	3 458
7	6	3	82.8	4 657	8 756
7	10	3	337	2 467	7 863

符号表示如下：

> PN：代表问题.
> DN：代表维数.
> IN：迭代次数.
> TI：计算时间.
> FN：函数计算个数.
> GN：梯度计算个数.

下面列出具体计算结果. 主要的迭代结果都集中在表 1 至表 7 每个函数的列表中. 用到的记号解释如下：

> x_k^0：第 k 步的初始点,
> k：第 k 个局部极小点的计数,
> x_k^*：第 k 个局部极小点,
> $f(x_k^*)$：第 k 个局部极小点的函数值.

表 1　函数 (i)

k	x_k^0	x_k^*	$f(x_k^*)$
1	$(-2, -1)$	$(-1.928\,303\,0, -0.806\,128\,3)$	$0.510\,688\,1$
2	$(0.484\,208\,2, 0.286\,498\,5)$	$(0.195\,666\,0, 0.719\,037\,8)$	$-0.989\,451\,8$

续　表

k	x_k^0	x_k^*	$f(x_k^*)$
3	$(-1.131\,439\,9E-02,$ $-0.686\,282\,1)$	$(-8.983\,376\,6e-02,$ $-0.712\,626\,2)$	$-1.031\,628$

表 2　函数 (ii)

k	x_k^0	x_k^*	$f(x_k^*)$
1	$(-2,-1)$	$(-1.018\,441,-1.639\,516)$	$742.129\,7$
2	$(-0.982\,801\,5,$ $-1.646\,444)$	$(-7.703\,313\,6E-04,$ $-1.000\,361)$	$3.000\,146$

表 3　函数(iii)

k	x_k^0	x_k^*	$f(x_k^*)$
1	$(-2,1)$	$(-2.000\,149,$ $4.999\,167\,5E-02)$	$2.499\,256\,1E-03$
2	$(-1.994\,570,$ $1.319\,889\,9E-02)$	$(-1.994\,574,$ $1.319\,626\,0E-02)$	$2.912\,506\,5E-04$
3	$(-7.205\,211\,1E-03,$ $-9.017\,001\,8E-03)$	$(-7.199\,509\,1E-03,$ $-9.015\,198\,8E-03)$	$2.871\,155\,4E-04$

表 4　函数 (iv)

k	x_k^0	x_k^*	$f(x_k^*)$
1	$(0.8,0.8)$	$(0.693\,889\,5,0.693\,889\,5)$	$-1.031\,207$
2	$(0.404\,027\,6,-0.313\,120\,2)$	$(0.346\,977\,7,-0.346\,893\,7)$	$-1.757\,801$
3	$(-0.323\,926\,4,$ $-1.350\,570\,5E-02)$	$(-0.346\,880\,3,$ $-2.470\,872\,3E-05)$	$-1.878\,900$
4	$(-1.173\,796\,6E-02,$ $1.993\,809\,3E-02)$	$(-3.179\,371\,2E-05,$ $5.467\,972\,4E-05)$	$-1.999\,999$

表 5　函数（v）

k	x_k^0	x_k^*	$f(x_k^*)$
1	(0, 0)	(4.156 207 3e-02, −9.480 731 2e-02)	0.517 455 5
2	(2.693 728, −0.926 118 0)	(0.552 726 5, −0.103 942 3)	3.322 207 9E-02
3	(4.795 631 0E-02, 0.296 607 0)	(0.230 017 2, 0.554 120 6)	3.938 106 4E-03
4	(−0.911 648 2, 0.805 844 5)	(0.102 525 0, 0.300 503 6)	5.370 603 5E-08

表 6　函数（vi）

k	x_k^0	x_k^*	$f(x_k^*)$
1	(1, 1)	(8.329 842, 8.329 842)	5.394 806 5e-07
2	(0.220 360 8, −1.386 927)	(−0.800 296 4, −1.425 057)	−186.730 9

表 7　函数（vii）

k	x_k^0	x_k^*	$f(x_k^*)$
1	(3, 3, 3, 3, 3, 3, 3, 3, 3, 3)	(−0.935 795 2, 1.997 097, 1.989 804, 1.989 650, 1.989 647, 1.989 646, 1.989 646, 1.989 646, 1.989 649, 1.989 754)	4.104 241
2	(−1.088 031, 1.002 956, 2.383 796, 1.024 900, 1.238 488, 1.795 130, 1.931 804, 1.791 897, 0.999 400 9, 1.167 541)	(−0.935 837 3, 1.001 203, 1.020 274, 1.000 000, 1.000 008, 1.000 007, 1.000 002, 1.134 899, 1.000 000, 1.002 461)	1.309 247

k	x_k^0	x_k^*	$f(x_k^*)$
3	$(-1.351\,654\,5E\text{-}02, 1.001\,229,$ $0.711\,046\,1, 1.619\,901,$ $1.022\,539, 1.198\,621,$ $0.742\,281\,7, 1.045\,286,$ $0.616\,074\,8, 1.717\,487)$	$(\ 1.000\,534, 0.998\,770\,4,$ $1.000\,245, 1.007\,303,$ $0.998\,412\,1, 0.998\,797\,7,$ $0.998\,315\,9, 0.999\,994\,5,$ $1.000\,466, 1.011\,805)$	$7.217\,459\,6E\text{-}05$

　　以上数值试验结果显示,本章的理论是正确的,具有打洞函数形式的变换函数应用于求解全局问题,计算的循环次数较少.

第三章 求解无约束全局优化问题的几个变换函数及其性质

§3.1 问题及假设

由于科学、经济、工程等领域的进步，很多问题涉及全局最优问题。因此，关于多极值的非线性全局最优化问题引起人们的高度注意。自二十世纪七十年代起很多求解全局最优问题的理论和算法得到了长足的发展。见 Dixon and Szegö(1975)[10]；Horst, Pardalos and Thoai (1995)[32] 等。由 Levy and Montalvo (1985)[47]提出的打洞函数法和由 Ge and Qin (1987)[17]提出的填充函数法是两个用于求解连续函数的全局最优化问题的方法，由于它们的实用性而得到了广泛的认可。

我们考虑无约束问题

$$\min \{f(x): x \in \Omega\} \tag{3.1.1}$$

其中 $f: R^n \to R$，$\Omega \subset R^n$ 是一个充分大的闭域。

对目标函数 $f(x)$ 有以下假设：

假设 3.1.1 $f(x)$ 在 R^n 连续可微，并且存在 $K > 0$ 使得 $\| \nabla f(x) \| \leqslant K$，$\forall x \in R^n$；

假设 3.1.2 $f(x) \to \infty$，当 $\| x \| \to \infty$，即 $f(x)$ 是一个强制函数；

假设 3.1.3 $f(x)$ 有有限个不同的极小值。

由于假设 3.1.2，可以认为 $\min \{f(x): x \in R^n\}$ 的所有极小点均在适当大的有界闭区域 Ω 内部。所以只需考虑 $f(x)$ 在 Ω 上的整体最小解,即问题(3.1.1),并且其所有极小点均是 Ω 的内点.

设问题 (3.1.1)有很多局部极小点,我们要求问题的全局最小点. 如果我们已经找到了一个局部极小 x_1^*,但它不是全局最小,我们可以用一个变换函数使迭代点列离开 x_1^* 所在的盆谷,找到 x' 使得 $f(x') \leqslant f(x_1^*)$. x' 所在的盆谷低于 x_1^* 所在的盆谷. 然后以 x' 为起点找出更好的局部极小点 x_2^*.

不论是用打洞函数法还是填充函数法求解问题(3.1.1),算法主要包括两个阶段.

第一阶段. 用一般的局部极小化算法,如拟牛顿法、共轭梯度法等找出目标函数 $f(x)$ 的一个局部极小点 x^*. 如果 x^* 不是全局最小点,算法进入第二阶段.

第二阶段. 其思想是定义一个变换函数——打洞函数或填充函数——利用它找出点 $\bar{x} \neq x^*$ 使得 $f(\bar{x}) < f(x^*)$. 几何上,点 $(\bar{x}, f(\bar{x}))$ 所在的盆谷比点 $(x^*, f(x^*))$ 所在的盆谷低. 然后以 $\bar{x}$ 为初始点,返回第一阶段.

两个阶段交替进行,直到找出整体最优解.

本章给出了几个简单、易于计算且函数性态较好的变换函数.

文章的以下部分设 x^* 是 $f(x)$ 的局部极小点但不是全局最优解. 而 x_G^* 是 $f(x)$ 在 Ω 上的一个全局最优解. 又令参数 r 满足:

$$0 < r < f(x^*) - f(x_G^*) \tag{3.1.2}$$

由(3.1.2)式可知,r 是依赖于问题的参数.

我们在区域 Ω 上定义两个子集:

$$S_1 = \{x: f(x) \geqslant f(x^*), x \in \Omega \backslash x^*\},$$

$$S_2 = \{x: f(x) < f(x^*), x \in \Omega\}.$$

即 $\Omega = S_1 \bigcup S_2 \bigcup \{x^*\}$.

§3.2 M-函数及其性质

设 M 是一个常数并且 $M \geqslant K$.

定义简单变换函数，M-函数：

$$P(x, x^*, r) = \frac{f(x) - f(x^*) + r}{r + M \| x - x^* \|} \qquad (3.2.1)$$

根据 $P(x, x^*, r)$ 的表达式，易知如果存在 $\bar{x} \neq x^*$ 使得 $P(\bar{x}, x^*, r) = 0$，则 $f(\bar{x}) < f(x^*)$. 即函数 $P(x, x^*, r)$ 具有打洞的特性，因而是一个变形打洞函数. 下面说明 $P(x, x^*, r)$ 亦有填充函数的某些特性.

函数 $P(x, x^*, r)$ 的梯度为：

$$\nabla P(x, x^*, r) = \frac{1}{r + M \| x - x^* \|} [\nabla f(x) -$$

$$\frac{M(f(x) - f(x^*) + r)}{(r + M \| x - x^* \|)} \cdot \frac{x - x^*}{\| x - x^* \|}]$$

$$= \frac{1}{r + M \| x - x^* \|} [\nabla f(x) -$$

$$M P(x, x^*, r) e_x]. \qquad (3.2.2)$$

显然 $\nabla P(x, x^*, r)$ 在 $\Omega \backslash \{x^*\}$ 上连续. 由下面的定理知 x^* 是 $P(x, x^*, r)$ 的极大点，因而在点 x^* 的附近 $P(x, x^*, r)$ 是凹函数，并且在 x^* 处 $P(x, x^*, r)$ 的次微分存在.

定理 3.2.1　如果假设 3.1.1 成立并且 $M \geqslant K$，则对任意 $x \neq x^*$，有 $P(x, x^*, r) \leqslant P(x^*, x^*, r)$ 成立.

证明　显然，$P(x^*, x^*, r) = 1$.

由假设 3.1.1，M 的取法和微分中值定理，可得

$$| P(x, x^*, r) | = \frac{| f(x) - f(x^*) + r |}{r + M \| x - x^* \|}$$

$$\leqslant \frac{|f(x)-f(x^*)|+r}{r+M\|x-x^*\|} \leqslant \frac{M\|x-x^*\|+r}{r+M\|x-x^*\|}=1,$$

即 $P(x, x^*, r) \leqslant |P(x, x^*, r)| \leqslant P(x^*, x^*, r), \forall x \neq x^*$.

定理 3.2.1 说明原问题的极小点 x^* 是变换函数 $P(x, x^*, r)$ 的全局极大点.

定理 3.2.2 (a) 设 x^* 是问题 (3.1.1)的局部极小点, 记

$$e_x = \frac{x-x^*}{\|x-x^*\|} \tag{3.2.3}$$

对任意 $x \in S_1$, 如果 $\nabla f(x) \neq MP(x, x^*, r)e_x$, 则 $\nabla P(x, x^*, r) \neq 0$.

(b) 如果 x^* 是问题 (3.1.1) 的一个局部极小点但不是全局最小点, 则函数 $P(x, x^*, r)$ 在集合 S_2 上有一个全局极小点.

证明 (a) 由(3.2.2)式知, $\nabla P(x, x^*, r) \neq 0$.

(b) 因为 x^* 不是 $f(x)$ 的全局极小, 所以 S_2 非空. 根据 $P(x, x^*, r)$ 的定义和(3.1.2)式, 我们有 $P(x_G^*, x^*, r) < 0$. 由假设 3.1.1 和 (3.2.2)式知, 函数 $P(x, x^*, r)$ 在区域 Ω 上连续, 因此在 Ω 有一全局极小点 $\bar{x}$, 并且有

$$P(\bar{x}, x^*, r) \leqslant P(x_G^*, x^*, r) < 0.$$

由此可得 $f(\bar{x})-f(x^*)+r<0$, 或 $f(\bar{x})<f(x^*)$, 即 $\bar{x} \in S_2$.

由定理 3.2.1 和定理 3.2.2 知, 当 $\nabla f(x) \neq MP(x, x^*, r)e_x$ 时, 函数 $P(x, x^*, r)$ 具有填充函数的性质. 在实际计算中, 当 $x \in S_1$ 时, 若出现 $\nabla f(x)=MP(x, x^*, r)e_x$ 的情况, 修正 M 的值即可.

若 $\bar{x}$ 是 Ω 的内点, 则有

推论 3.2.1 在定理 3.2.2 的条件下, 有 $x \in S_2$ 使得 $\nabla P(x, x^*, r) = 0$.

算例 3.2.1

$$f(x) = [1-2x_2+c\sin(4\pi x_2)-x_1]^2 + [x_2-0.5\sin(2\pi x_1)]^2$$
$$-10 \leqslant x_1, x_2 \leqslant 10$$

这里 $c=0.2$ 或 0.5.

不论 c 取何值,$f(x)$ 的全局最小值 $f(x_G^*)=0$,其中 x_G^* 是全局最优点. $f(x)$ 有很多全局解. 表 8 列出 $c=0.2$ 时,取不同的初始点的计算结果. 表中 x_0 为初始点. 图 1 是 $(0,0)$ 点附近的图形.

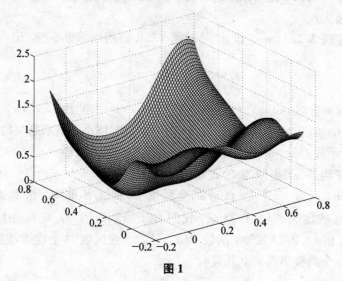

图 1

表 8　算例 3.2.1

x_0	x_G^*	$f(x_G^*)$
$(0,0)$	$(1.000\,005\,9,\ 0.000\,019\,5)$	$1.671\,270\,9 * 10^{-11}$
$(5,0)$	$(0.999\,975\,7,\ -0.000\,065\,3)$	$2.044\,264\,2 * 10^{-10}$
$(5,4)$	$(1.878\,455\,4,\ -0.345\,866\,7)$	$5.758\,306\,8 * 10^{-9}$
$(-5,5)$	$(0.409\,089\,3,\ 0.270\,293\,8)$	$1.783\,210\,5 * 10^{-8}$

§3.3　α -函数及其性质

本节给出另一个简单函数,α -函数:

$$Q(x, x^*, r, \alpha) = \frac{f(x) - f(x^*) + r}{[r + \| x - x^* \|]^\alpha} \qquad (3.3.1)$$

其中 $\alpha > 0$, r 和 x^* 的定义与 M-函数相同.

易见, $Q(x, x^*, r, \alpha)$ 亦是一个变形打洞函数.

函数 $Q(x, x^*, r, \alpha)$ 的梯度为:

$$\nabla Q(x, x^*, r, \alpha) = \frac{1}{[r + \| x - x^* \|]^\alpha} \{ \nabla f(x) -$$

$$\frac{\alpha[f(x) - f(x^*) + r]}{r + \| x - x^* \|} \cdot \frac{(x - x^*)}{\| x - x^* \|} \}$$

$$(3.3.2)$$

由 $f(x)$ 的连续可微性知, $\nabla Q(x, x^*, r, \alpha)$ 在 $\Omega / \{x^*\}$ 上连续. 在点 x^* 处 $Q(x, x^*, r, \alpha)$ 的次微分存在.

记

$$\overline{D} = \max_{x \in \Omega} \| x - x^* \|.$$

有如下定理.

定理 3.3.1 在假设 3.1.1 下,当

$$\alpha > \frac{K(r + \overline{D})}{r}$$

对所有 $x \in S_1$ 有

$$(x - x^*)^T \nabla Q(x, x^*, r, \alpha) < 0 \qquad (3.3.3)$$

证明 沿用 (3.2.3) 的记号. 对 $f(x) \geqslant f(x^*)$, 即 $x \in S_1$, 由假设 3.1.1 和 (3.3.2) 式得

$$e_x^T \nabla Q(x, x^*, r, \alpha)$$

$$= \frac{1}{[r + \| x - x^* \|]^\alpha} \left[e_x^T \nabla f(x) - \frac{\alpha(f(x) - f(x^*) + r)}{r + \| x - x^* \|} \right]$$

$$\leqslant \frac{1}{[r+\|x-x^*\|]^\alpha}\left(\|\nabla f(x)\| - \frac{\alpha r}{r+\|x-x^*\|}\right)$$

$$\leqslant \frac{1}{[r+\|x-x^*\|]^\alpha}\left(K - \frac{\alpha r}{r+\overline{D}}\right),$$

所以 $\alpha > \dfrac{K(r+\overline{D})}{r}$ 时，$e_x^{\mathrm{T}} \nabla Q(x, x^*, r, \alpha) < 0$，即(3.3.3)式成立.

定理 3.3.1 说明对任意 $x \in S_1$，即 $f(x) \geqslant f(x^*)$，当 α 适当大后，$x-x^*$ 是函数 $Q(x, x^*, r, \alpha)$ 在 x 点处的严格下降方向.

推论 3.3.1 在定理 3.3.1 的条件下，α-函数 $Q(x, x^*, r, \alpha)$ 在 S_1 上没有稳定点.

定理 3.3.2 设 $x_1, x_2 \in S_1$ 并且有 $\varepsilon > 0$ 使得 $\|x_1-x^*\| \geqslant \|x_2-x^*\|+\varepsilon$，则当

$$\alpha > \frac{\ln r - \ln(K\overline{D}+r)}{\ln(\overline{D}+r-\varepsilon) - \ln(\overline{D}+r)}$$

时 $Q(x_1, x^*, r, \alpha) < Q(x_2, x^*, r, \alpha)$.

证明 因为 $\|x_1-x^*\| \geqslant \|x_2-x^*\|+\varepsilon$，所以

$$\frac{r+\|x_2-x^*\|}{r+\|x_1-x^*\|} \leqslant 1 - \frac{\varepsilon}{\overline{D}+r}.$$

又由假设 $\|\nabla f(x)\| \leqslant K$ 得

$$\frac{f(x_2)-f(x^*)+r}{f(x_1)-f(x^*)+r} \geqslant \frac{r}{K\overline{D}+r}.$$

故当 $\alpha > \dfrac{\ln r - \ln(K\overline{D}+r)}{\ln(\overline{D}+r-\varepsilon) - \ln(\overline{D}+r)}$ 时，有

$$\frac{r}{K\overline{D}+r} > \left(1 - \frac{\varepsilon}{\overline{D}+r}\right)^\alpha,$$

从而

$$\frac{f(x_2) - f(x^*) + r}{f(x_1) - f(x^*) + r} > \left(\frac{r + \parallel x_2 - x^* \parallel}{r + \parallel x_1 - x^* \parallel}\right)^\alpha,$$

即 $Q(x_1, x^*, r, \alpha) < Q(x_2, x^*, r, \alpha)$.

由以上两个定理可得:

推论 3.3.2 对充分大的 α, x^* 是函数 $Q(x, x^*, r, \alpha)$ 的全局极大点,且极大值 $Q(x^*, x^*, r, \alpha) = 1$.

下面的定理说明存在 $x \in S_2$ 使得 $Q(x, x^*, r, \alpha)$ 沿 $x - x^*$ 的方向导数为零.

定理 3.3.3 若 x^* 不是全局最优解,则当 α 充分大时,有 $x \in S_2$ 使得

$$(x - x^*)^{\mathrm{T}} \nabla Q(x, x^*, r, \alpha) = 0.$$

证明 因为 x^* 不是全局最优解,所以 S_2 非空. 令

$$\phi(x) = e_x^{\mathrm{T}} \nabla f(x) - \frac{\alpha(f(x) - f(x^*) + r)}{r + \parallel x - x^* \parallel} \qquad (3.3.4)$$

如果要证

$$(x - x^*)^{\mathrm{T}} \nabla Q(x, x^*, r, \alpha)$$

$$= \frac{\parallel x - x^* \parallel}{(r + \parallel x - x^* \parallel)^\alpha} \left[e_x^{\mathrm{T}} \nabla f(x) - \frac{\alpha(f(x) - f(x^*) + r)}{r + \parallel x - x^* \parallel} \right] = 0,$$

只需证 $\phi(x) = 0$.

由 r 的定义和 $\nabla f(x_G^*) = 0$ 知 $\phi(x_G^*) > 0$. 注意到定理 3.3.1 的证明,易知当 $x \in S_1$ 并且 α 充分大时 $\phi(x) < 0$. 在点 x^* 处,因为 $\nabla f(x^*) = 0$,所以由 $x \to x^*$ 时 $(x - x^*)^{\mathrm{T}} \nabla f(x) = o(\parallel x - x^* \parallel)$ 可得 $\phi(x^*) = -\alpha < 0$. 因此根据 $\phi(x)$ 的连续性,有 $\bar{x} \neq x^*$ 存在使得 $\phi(\bar{x}) = 0$. 由定理 3.3.1 知 $\bar{x} \notin S_1$,所以 $\bar{x} \in S_2$.

类似于定理 3.2.2(b) 的证明,易知下面的定理是成立的.

定理 3.3.4 如果 x^* 是问题 (3.3.1) 的一个局部极小点但不是全局最小点,则函数 $Q(x, x^*, r, \alpha)$ 在集合 S_2 上有一个全局极小点.

算例 3.3.1 *Goldstein and Price* 函数 [20]:

$$f(x) = [1 + (x_1 + x_2 + 1)^2 (19 - 14x_1 + 3x_1^2 -$$

$$14x_2 + 6x_1 x_2 + 3x_2^2)] \times [30 + (2x_1 - 3x_2)^2$$

$$(18 - 32x_1 + 12x_1^2 + 48x_2 - 36x_1 x_2 + 27x_2^2)],$$

$$-3 \leqslant x_1, x_2 \leqslant 3$$

取初始点为 $x^0 = (1, 1)$,得三个局部极小点 $x_1^* = (1.200\,004\,7,$ $0.799\,978\,5)$, $x_2^* = (1.800\,034\,2, 0.200\,021\,9)$, $x_3^* = (0.000\,004\,0,$ $-1.000\,023\,0)$,此问题有唯一全局解 $x_G^* = (0, -1)$, $f(x_G^*) = 3$,故 x_3^* 为全局解. 见图 2.

取初始点为 $x^0 = (-2, -1)$,一次循环即可达全局解 $x^* = (0.000\,047\,1, -0.999\,988\,8)$.

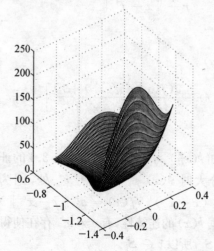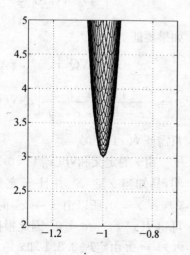

图 2

§3.4 梯度型变形辅助函数

求解全局最优化问题的辅助函数多数是 $f(x)$ 的复合函数，较少用到其梯度 $\nabla f(x)$ 的复合. 文[81]是少数由 $\nabla f(x)$ 构成辅助函数的方法之一. 此文提出由两个阶段: 动态优化阶段和动态打洞阶段组成的动态打洞函数方法. 在动态优化阶段求解动态系统

$$\frac{\mathrm{d}x_i}{\mathrm{d}t} = -\frac{\partial f(x)}{\partial x_i}, \, i = 1, 2, \cdots, n$$

的平衡点，即局部最优解 x^*. 在动态打洞阶段不是求解打洞函数(2.1.2)的零点，而是用 $\nabla f(x)$ 的复合函数构造动态打洞系统:

$$\frac{\mathrm{d}x_i}{\mathrm{d}t} = -\frac{\dfrac{\partial f}{\partial x_i}}{\parallel x - x^* \parallel^\lambda} - k\hat{f}(f(x) - f(x^*))\frac{\partial f}{\partial x_i}, \, i = 1, 2, \cdots, n$$

$$(3.4.1)$$

其中

$$\hat{f}(z) = \begin{cases} z, \, z > 0 \\ 0, \, z \leqslant 0. \end{cases}$$

k 是罚参数. 当动态系统(3.4.1)收敛到平衡状态时，相当于在约束 $f(x) - f(x^*) \leqslant 0$ 下求打洞函数的零点，亦即完成了打洞阶段的任务，所以系统(3.4.1)的平衡点 $\bar{x}$ 在比 x^* 低的盆谷里.

现在我们从不同于动态打洞系统的观点提出由 $f(x)$ 的梯度组成的变换函数. 大家知道在 $f(x)$ 的极小点 $\bar{x}$，或任一驻点 $\bar{x}$ 处有 $\nabla f(\bar{x}) = 0$，并且在相邻的驻点(包括极大值点，极小值点，鞍点)之间 $\parallel \nabla f(x) \parallel$ 不为零. 基于此给出如下单参数变换函数:

$$G(x, x^0, r) = \frac{1}{2}\parallel \nabla f(x) \parallel^2 + \ln\left(1 + \exp\left\{\frac{f(x) - f(x^0)}{r}\right\}\right)$$

$$(3.4.2)$$

这里 x^0 是闭区域 Ω 内的一个点；$0 < r < 1$.

本节将假设 3.1.1 改为

假设 3.4.1 $f(x)$ 有二阶连续偏导数，其 *Hessian* 矩阵记为 $H(x)$.
记

$$\zeta(x) = \frac{f(x) - f(x^0)}{r}.$$

$G(x, x^0)$ 关于 x 的梯度为：

$$\nabla G(x, x^0, r) = \left(H(x) + \frac{1}{r} \frac{e^{\zeta(x)}}{1 + e^{\zeta(x)}} E \right) \nabla f(x) \qquad (3.4.3)$$

这里 E 为单位矩阵.

定理 3.4.1 在 x 处，*Hessian* 阵 $H(x)$ 的特征值为 $\lambda_i(x)(i = 1, 2, \cdots, n)$. 如果 $\forall x \in S_1(x^0) = \{x \mid f(x) \geqslant f(x^0), x \in \Omega\}$，$\| H(x) \|$ 有界，令

$$\bar{\lambda} = \max_{x \in S_1(x^0)} \{ \max_{1 \leqslant i \leqslant n} | \lambda_i(x) | \},$$

则当

$$0 < r < \frac{1}{2\bar{\lambda}}$$

以后，$(3.4.3)$ 式中的矩阵 $H(x) + \dfrac{1}{r} \dfrac{e^{\zeta(x)}}{1 + e^{\zeta(x)}} E$ 为 $S_1(x^0)$ 上的对称正定矩阵.

证明 显然 $H(x) + \dfrac{1}{r} \dfrac{e^{\zeta(x)}}{1 + e^{\zeta(x)}} E$ 是对称矩阵.

因为在点 x 处，存在正交矩阵 $R = R(x)$ 使得 $H(x) = R^T \Lambda(x) R$，这里 $\Lambda(x) = \mathrm{diag}(\lambda_1(x), \cdots, \lambda_n(x))$，所以有

$$H(x) + \frac{1}{r} \frac{e^{\zeta(x)}}{1 + e^{\zeta(x)}} E = R^T \left(\Lambda(x) + \frac{1}{r} \frac{e^{\zeta(x)}}{1 + e^{\zeta(x)}} E \right) R.$$

因为 $x \in S_1(x^0)$ 时 $\zeta(x) \geqslant 0$，进而有

$$\frac{1}{2r} \leqslant \frac{1}{r} \frac{\mathrm{e}^{\zeta(x)}}{1+\mathrm{e}^{\zeta(x)}} \leqslant \frac{1}{r},$$

因此当 $0 < r < \dfrac{1}{2\bar{\lambda}}$，矩阵 $\Lambda(x) + \dfrac{1}{r} \dfrac{\mathrm{e}^{\zeta(x)}}{1+\mathrm{e}^{\zeta(x)}} E$ 正定，从而 $H(x) +$

$\dfrac{1}{r} \dfrac{\mathrm{e}^{\zeta(x)}}{1+\mathrm{e}^{\zeta(x)}} E$ 正定.

注意到(3.4.3)式，定理 3.4.1 说明，在区域 $S_1(x^0)$ 上对适当的 r，$f(x)$ 和 $G(x, x^0)$ 有相同的下降方向. 实际计算中 r 是一个需要人工调节的参数.

我们考虑辅助问题：

$$\min_{x \in \Omega} G(x, x^0, r) \tag{3.4.4}$$

定理 3.4.2 如果 x^0 不是原问题(3.1.1)的全局解，则当 $r \to 0$ 时辅助问题(3.4.4)的最优值是零.

证明 因为 x^0 不是原问题(3.1.1)的全局解，所以 $S_2(x^0) = \{x \mid f(x) < f(x^0), x \in \Omega\} \neq \varnothing$，问题 (3.1.1) 的全局解 $x_G^* \in S_2(x^0)$. 由 (3.4.2) 式知 $G(x, x^0, r) > 0$，$\forall x \in \Omega$. 又根据假设 3.1.2 知 $\nabla f(x_G^*) = 0$，从而 $\lim\limits_{r \to 0} G(x_G^*, x^0, r) = 0$. 所以辅助问题(3.4.4)的解 $\bar{x}$ 满足 $0 \leqslant \lim\limits_{r \to 0} G(\bar{x}, x^0, r) \leqslant \lim\limits_{r \to 0} G(x_G^*, x^0, r) = 0$. 得证.

由定理 3.4.2 及证明可得

推论 3.4.1 如果 x^0 不是原问题(3.1.1)的全局解，则辅助问题(3.4.4)在区域 $S_1(x^0)$ 上没有最优解.

证明 注意到当 $x \in S_1(x^0)$ 时，$G(x, x^0, r) \geqslant \ln 2$，由定理 3.4.2 即知结论成立.

由定理 3.4.2 又有

定理 3.4.3 当 $r \to 0$ 时，原问题(3.1.1)的全局最优解必是辅助问题(3.4.4)的全局解.

　　由以上几个结论知道，相对于打洞函数法和填充函数法，用变换函数 $G(x, x^0, r)$ 求解问题 (3.1.1) 的全局解的优点是不需要极小化和打洞两个阶段交替进行，只需求解问题 (3.4.4)，在求解的过程中调节 r 的值，变更 x^0.

　　可用下降法求解问题 (3.4.4)，得到最优解 $\bar{x}$，必是原问题 (3.1.1) 的稳定点，但可能不是其极小点. 然而这并不影响算法的有效性. 下面是理论算法.

算法

　　1. $\forall x_0 \in \Omega$；$0 < r < 1$；充分小的正数 ε 为控制参数.

　　2. 求解问题 (3.4.4).

　　3. 如果问题 (3.4.4) 有解 $\bar{x} \neq x^0$，且 $G(\bar{x}, x^0, r) = 0$，则令 $x^0 = \bar{x}$，转 2. 否则转 4.

　　4. 如果 (3.4.4) 的最优值为 $\ln 2$，且 $r < \varepsilon$ 则停，问题 (3.4.4) 的当前解 $\bar{x}$ 即为原问题 (3.1.1) 的最优解. 否则令 $r = \dfrac{1}{2} r$. 转 2.

第四章　求解约束全局最优问题的变换函数及其算法

本章主要包括两部分内容. 第一节到第三节讨论带线性不等式约束的非线性规划问题全局解的算法及性质. 第四节讨论一般的带非线性不等式约束的非线性规划问题全局解的算法及变换函数的性质.

§4.1　线性约束非凸规划问题及其假设

对于求解带有线性约束的非线性规划问题的局部解, 已有很多算法. 如[66, 83, 73]等是用次最优化方法求解此类问题. 由于这些方法考虑的都是局部问题, 所以要得到全局解必要有"凸"性条件. 本章考虑去掉目标函数的凸性要求, 将用于无约束全局最优问题的思想方法拓展到求解带有线性约束的非线性规划问题的全局最优问题.

考虑问题

$$\min \{f(x) : x \in X\} \tag{4.1.1}$$

其中 $f(x) \in C^1 : R^n \to R$, $X = \{x \in R^n : Ax \leqslant b\}$ 为可行域, A 为 $m \times n$ 阶矩阵, $b = (b_1, b_2, \cdots, b_m)^T \in R^m$.

为方便起见, 下面给出一些记号.

记指标集 $J = \{1, 2, \cdots, m\}$.

矩阵 A 的第 i 行为 a_i^T.

$J(x) = \{i : a_i^T x = b_i, i \in J\}$ 为紧约束指标集, 如果 $J(x) = \varnothing$ 则问题退化为无约束问题, 所以本章假设 $J(x) \neq \varnothing$, $J(x)$ 的元素个

数为 $|J(x)| = S$.

$A_x = \{a_i^T : i \in J(x)\}$;

$b_x = (b_i : i \in J(x))^T \in R^S$;

超平面 $H_J = \{x \mid A_x x = b_x\}$.

$S_1(\bar{x}) = \{x \mid f(x) \geqslant f(\bar{x}), x \neq \bar{x}\}$;

$S_2(\bar{x}) = \{x \mid f(x) < f(\bar{x})\}$.

$L(P)$ 代表问题(4.1.1) 的局部极小点集合;

$G(P)$ 为问题(4.1.1) 的全局最优解集合.

本章总是认为以下假设成立:

假设 4.1.1 $f(x)$ 只有有限个不同的极小值;

假设 4.1.2 $f(x)$ 的梯度向量 $\nabla f(x)$ 在直径充分大的闭集 Ω 上连续;

假设 4.1.3 子矩阵 A_x 的行向量组线性无关.

由假设 4.1.1 知,有直径充分大的有界闭箱 Ω 使得 $L(P) \subset \Omega$,即问题(4.1.1)的局部极小点全部在 Ω 的内部. 我们只需考虑问题(4.1.1)在有界闭区域 $X \cap \Omega$ 上的全局解即可. 因此我们可以认为 X 是有界闭集,且算法中的 x 总是取自 Ω.

§4.2 算法

下面给出算法:

步骤 1 在可行域 X 内取一点 x^0.

步骤 2 以 x^0 为起点,求解问题(4.1.1),得局部极小点 $\bar{x}^1$.

步骤 3 用下面子步骤求解子问题

$$\min f(x)$$

$$s.t. \ a_i^T x = b_i, i \in J(\bar{x}^1), x \in \Omega \qquad (4.2.1)$$

的全局解 $\bar{x}^*$.

步骤 3.1 记 $\bar{A} = A_{\bar{x}^1}$, $\bar{b} = b_{\bar{x}^1}$, I 为 $n \times n$ 阶单位矩阵. 令 $P =$

$I-\bar{A}^{\mathrm{T}}(\bar{A}\,\bar{A}^{\mathrm{T}})^{-1}\,\bar{A}$. $k=1$.

步骤 3.2 在点 $\bar{x}^k$ 处构造变换函数 $T(x,\bar{x}^k)$，求解辅助问题

$$\min T(x,\bar{x}^k)$$
$$s.t.\bar{A}x=\bar{b} \tag{4.2.2}$$

如果辅助问题(4.2.2)没有解，则 $\bar{x}^*=\bar{x}^k$，转步骤 4；否则

步骤 3.3 设辅助问题(4.2.2)的解为 $\bar{x}'$. 若 $\bar{x}'$ 满足 $A\bar{x}'\leqslant b$，转步骤 3.4；否则 $\bar{x}^*=\bar{x}^k$，转步骤 4.

步骤 3.4 以 $\bar{x}'$ 为起点，在点 x 处以 $d_f(x)=-P\nabla f(x)$ 为下降方向，求解子问题(4.2.1)，得局部极小点 $\bar{x}^{k+1}$. 如果 $A\bar{x}^{k+1}\leqslant b$，令 $k=k+1$，转步骤 3.2；否则令 $x^0=\bar{x}'$，转步骤 2.

步骤 4 令

$$\lambda_J=-(\bar{A}\,\bar{A}^{\mathrm{T}})^{-1}\,\bar{A}\,\nabla f(\bar{x}^*)=(\lambda_j,j\in J(\bar{x}^1))^{\mathrm{T}} \tag{4.2.3}$$

(i) 如果存在 $\lambda_j<0$，则令 $x^0=\bar{x}^*$，转步骤 2.

(ii) 如果 $\lambda_J\geqslant 0$ 且有 $j\in J\backslash J(\bar{x}^1)$，使得 $a_j^{\mathrm{T}}\bar{x}^*=b_j$，则令 $\bar{x}^1=\bar{x}^*$ 及 $J(\bar{x}^1)=J(\bar{x}^1)\bigcup\{i\}$，转步骤 3.

(iii) 如果 $\lambda_J\geqslant 0$ 但对所有 $j\in J\backslash J(\bar{x}^1)$，有 $a_j^{\mathrm{T}}\bar{x}^*<b_j$，则转步骤 5.

步骤 5 在点 $\bar{x}^*$ 处构造求解约束全局优化问题的变换函数 $Q(x,\bar{x}^*,a_1^{\mathrm{T}}x-b_1,\cdots,a_m^{\mathrm{T}}x-b_m)$，求解问题

$$\min Q(x,\bar{x}^*,a_1^{\mathrm{T}}x-b_1,\cdots,a_m^{\mathrm{T}}x-b_m) \tag{4.2.4}$$

若问题(4.2.4)无解，则 $\bar{x}^*$ 为问题(4.1.1)的全局解，算法终止；若问题(4.2.4)有解 $\hat{x}$，则令 $x^0=\hat{x}$，转步骤 2.

§4.3 算法的性质

记子空间

$$V_0 = \{z \mid \bar{A}z = 0\},$$

则

$$V_0^\perp = \{x \mid x = \bar{A}^\mathrm{T} v, \, v \in R^S\}.$$

根据假设 4.1.3, 矩阵 $\bar{A}$ 行满秩. 设 $\bar{A} = (B \quad N)$, 其中矩阵 B 是 $S \times S$ 阶可逆矩阵. 易知矩阵

$$\begin{pmatrix} -B^{-1}N \\ I_{n-S} \end{pmatrix}$$

的列向量组成 V_0 的一个基. 因此存在矩阵 Z 其列向量组成 V_0 的正交基, 所以

$$V_0 = \{z \mid z = Zt, \, t \in R^{n-S}\},$$

并且有 $\bar{A}Z = 0$ 和 $Z^\mathrm{T}Z = I_{(n-S) \times (n-S)}$. 事实上, 对矩阵 $\bar{A}^\mathrm{T}$ 进行 QR 分解:

$$\bar{A}^\mathrm{T} = (Y \quad Z) \begin{pmatrix} R \\ 0 \end{pmatrix}$$

其中 $Q = (Y \quad Z)$ 是正交矩阵. 易证子矩阵 Z 即为所求. 因此有下列引理.

引理 4.3.1 如果 $\bar{A}$ 行满秩, 则存在矩阵 Z 使得 $Z^\mathrm{T}Z = I_{(n-S) \times (n-S)}$, ZZ^T 是投影算子, 且

$$P = ZZ^\mathrm{T} = I - \bar{A}^\mathrm{T}(\bar{A}\,\bar{A}^\mathrm{T})^{-1}\,\bar{A} \tag{4.3.1}$$

证明 见参考文献[94, 66].

引理 4.3.2 矩阵 P 对称半正定, 且 $\bar{A}P = 0$.

证明 显然 $P^\mathrm{T} = P$. 注意到 $P^2 = P$, 所以 $\forall x \in R^n$ 有 $x^\mathrm{T}Px = \| Px \|^2 \geqslant 0$. 另外由 (4.3.1) 式可知 $\bar{A}P = 0$ 成立.

由引理 4.3.2 知, $\forall x \in R^n$ 有 $Px \in V_0$.

引理 4.3.3 $Px=0 \Leftrightarrow x \in V_0^\perp$.

证明 ($\Leftarrow$)设 $x \in V_0^\perp$,则有 $v \in R^s$ 使得 $x = \bar{A}^T v$,所以由引理 4.3.2,$Px = P^T \bar{A}^T v = (\bar{A}P)^T v = 0$.

($\Rightarrow$)由 $Px = 0$ 得 $x^T Px = 0$. 根据引理 4.3.1,$x^T ZZ^T x = 0$,即 $Z^T x = 0$. $\forall z \in V_0$,因为 $z^T x = t^T Z^T x = 0$,所以 $x \in V_0^\perp$.

如果令

$$\mathrm{d}(x) = -P \nabla f(x)$$

则有

定理 4.3.1 (1) 若 $d(x) \neq 0$,则 $\mathrm{d}(x)$ 是子问题(4.2.1)的可行下降方向.

(2) 存在 $\bar{x} \in H_J \bigcap \Omega$,使得 $\mathrm{d}(\bar{x}) = 0 \Leftrightarrow \bar{x}$ 是子问题(4.2.1)的 KKT 点.

证明 (1) $\forall x \in H_J \bigcap \Omega$,由引理 4.3.2 知:

$\bar{A} \mathrm{d}(x) = -\bar{A}P \nabla f(x) = 0$. 所以 $\mathrm{d}(x)$ 可行. 由于 $\mathrm{d}(x) = -P \nabla f(x) \neq 0$,所以 $\nabla f(x)^T \mathrm{d}(x) = -\nabla f(x)^T P \nabla f(x) < 0$,即 $\mathrm{d}(x)$ 是 $f(x)$ 在点 x 的下降方向.

(2) 由 $\mathrm{d}(x)$ 的定义和引理 4.3.3,$\mathrm{d}(\bar{x}) = 0 \Leftrightarrow P \nabla f(\bar{x}) = 0 \Leftrightarrow \nabla f(\bar{x}) \in V_0^\perp$,即有 $v \in R^s$ 使得 $\nabla f(\bar{x}) = \bar{A}^T v$.

在算法步骤 3.2 中,取变换函数

$$T(x, \bar{x}) = \frac{f(x) - f(\bar{x}) + r}{\| x - \bar{x} \|^\alpha} \qquad (4.3.2)$$

其中 $r > 0$ 且适当小;$\alpha \geqslant 1$;$\bar{x}$ 是问题(4.2.1)的局部极小点. 那么辅助问题(4.2.2)的下降方向可取为

$$\mathrm{d}_T = \mathrm{d}_f + \frac{\alpha(f(x) - f(\bar{x}) + r)}{\| x - \bar{x} \|^2}(x - \bar{x}) \qquad (4.3.3)$$

其中 $\mathrm{d}_f = -P \nabla f(x)$.

$T(x, \bar{x})$ 的梯度函数：

$$\nabla T(x, \bar{x}) = \frac{1}{\| x - \bar{x} \|^{\alpha}} \left[\nabla f(x) - \frac{\alpha(f(x) - f(\bar{x}) + r)}{\| x - \bar{x} \|^{2}} (x - \bar{x}) \right].$$

如果记 $\bar{A} = (B \quad N)$，其中矩阵 B 是 S 阶可逆方阵，相应的 $x = (x_B, x_N)^{\mathrm{T}}$. 由 $\bar{A}x = \bar{b}$ 可得 $x_B = B^{-1}\bar{b} - B^{-1}Nx_N$. 代入目标函数

$$f(x) = f(x_B, x_N) = f(B^{-1}\bar{b} - B^{-1}Nx_N, x_N),$$

记

$$\bar{f}(x_N) = f(B^{-1}\bar{b} - B^{-1}Nx_N, x_N) \tag{4.3.4}$$

$$\nabla f(x) = \begin{pmatrix} \nabla f_B(x) \\ \nabla f_N(x) \end{pmatrix}$$

则

$$\nabla \bar{f}(x_N) = -(B^{-1}N)^{\mathrm{T}} \nabla f_B(x) + \nabla f_N(x) \tag{4.3.5}$$

定理 4.3.2 设本章假设 4.1.1—4.1.3 成立，并记 $\bar{x}^k = \bar{x}$. 在超平面 $H = H_J = \{x \mid \bar{A}x = \bar{b}\}$ 与 Ω 的交集上，有以下结论成立：

(1) 若 $x \neq \bar{x}$ 且 $f(x) \geqslant f(\bar{x})$，则当 α 适当大，有 $(x - \bar{x})^{\mathrm{T}} \nabla T(x, \bar{x}) < 0$ 成立.

(2) 若 $\bar{x}$ 不是子问题 (4.2.1) 的全局极小，则必有适当的 α 和 r 使得存在 $\bar{x}' \in \{x \mid f(x) < f(\bar{x}), x \in H \bigcap \Omega\}$ 成为 $T(x, \bar{x})$ 在 $H \bigcap \Omega$ 上的极小点.

证明 (1) 设 $x \in H$，$x \neq \bar{x}$，且 $f(x) \geqslant f(\bar{x})$，则

$$(x - \bar{x})^{\mathrm{T}} \nabla T(x, \bar{x}) = \frac{1}{\| x - \bar{x} \|^{\alpha}} \big[(x - \bar{x})^{\mathrm{T}} \nabla f(x) -$$

$$\alpha (f(x) - f(\bar{x}) + r) \big]$$

$$\leqslant \frac{1}{\|x-\bar{x}\|^{\alpha}}\left[(x_N-\bar{x}_N)^{\mathrm{T}}\begin{pmatrix}-B^{-1}N\\I_{n-S}\end{pmatrix}^{\mathrm{T}}\begin{pmatrix}\nabla f_B(x)\\\nabla f_N(x)\end{pmatrix}-\alpha r\right]$$

$$=\frac{1}{\|x-\bar{x}\|^{\alpha}}\left[(x_N-\bar{x}_N)^{\mathrm{T}}\nabla\bar{f}(x_N)-\alpha r\right].$$

由假设 4.1.1 和 4.1.2 知, 存在 $M>0$, 使得 $\|(x_N-\bar{x}_N)^{\mathrm{T}}\nabla\bar{f}(x_N)\|\leqslant M$, 所以当

$$\alpha>\frac{M}{r}$$

时, 有 $(x-\bar{x})^{\mathrm{T}}\nabla T(x,\bar{x})<0$ 成立.

(2) 由于 $\bar{x}$ 不是问题(4.2.1)在 $H\bigcap\Omega$ 上的全局极小点, 所以集合 $\{x\mid f(x)<f(\bar{x}), x\in H\bigcap\Omega\}$ 非空.

由 $T(x,\bar{x})$ 的定义和(4.3.4)式,

$$T(x,\bar{x})=\frac{f(x)-f(\bar{x})+r}{\|x-\bar{x}\|^{\alpha}}$$

$$=\frac{\bar{f}(x_N)-\bar{f}(\bar{x}_N)+r}{\left\|\begin{pmatrix}-B^{-1}N\\I_{n-S}\end{pmatrix}(x_N-\bar{x}_N)\right\|^{\alpha}}$$

$$\triangleq\overline{T}(x_N,\bar{x}_N),$$

即辅助问题(4.2.2)是超平面 H 上的 $n-S$ 维无约束问题.

由假设 4.1.2 以及

$$\lim_{x_N\to\bar{x}_N}\overline{T}(x_N,\bar{x}_N)=+\infty,$$

即函数 $\overline{T}(x_N,\bar{x}_N)$ 只在 x_N 处无界, 我们知存在 $\bar{x}_N$ 的邻域 $O(\bar{x}_N)\subset H$, 使得 $\overline{T}(x_N,\bar{x}_N)$ 在 $H\backslash O(\bar{x}_N)$ 上连续. 因此函数 $T(x,\bar{x})$ 在有界闭区域 $H\bigcap\{\Omega\backslash O(\bar{x})\}$ 上连续, 必取得最小值, 令

$$\bar{x}' = arg \min_{x \in H \cap \{ \Omega \setminus \alpha(\bar{x}) \}} T(x, \bar{x}) \qquad (4.3.6)$$

$\bar{x}$ 和 $\bar{x}'$ 都在超平面 H 上.

另一方面,由定理假设 $f(\bar{x})$ 不是子问题(4.2.1)的全局极小值,而函数 $f(x)$ 在有界闭区域 $H \cap \Omega$ 上连续,必在此区域取得最小值 $f(x_H)$,其中 x_H 在超平面 H 上且是子问题(4.2.1)的全局极小. 取 $0 < r < f(\bar{x}) - f(x_H)$,由 $T(x, \bar{x})$ 的定义(4.3.2)式知,必有 $T(x_H, \bar{x}) < 0$.

由 $\bar{x}'$ 的定义(4.3.6)式知 $T(\bar{x}', \bar{x}) < T(x_H, \bar{x}) < 0$. 根据 (4.3.2)式,这意味着 $f(\bar{x}') < f(\bar{x})$,亦即 $\bar{x}' \in \{x \mid f(x) < f(\bar{x}), x \in H \cap \Omega \}$. 定理得证.

定理 4.3.2 说明在集合 $H \cap S_1(\bar{x})$ 上,$x - \bar{x}$ 是变换函数 $T(x, \bar{x})$ 的下降方向,因此 $T(x, \bar{x})$ 在其上没有稳定点. $T(x, \bar{x})$ 的稳定点 (如果存在的话)一定在集合 $H \cap S_2(\bar{x})$ 上. 故而对变换函数 $T(x, \bar{x})$ 用下降算法有如下定理成立.

定理 4.3.3 当变换函数为(4.3.2)式时,若 $x \in H \cap \Omega$ 且 $d_T \neq 0$,则 d_T 是辅助问题(4.2.2)的可行下降方向.

证明 根据 d_T 的表达式,并注意到引理 4.3.2 及 x 和 $\bar{x}$ 均满足 $\bar{A}x = \bar{b}$,有

$$\bar{A}d_T = -\bar{A}P\nabla f(x) + \frac{\alpha(f(x) - f(\bar{x}) + r)}{\| x - \bar{x} \|^2} \bar{A}(x - \bar{x}) = 0.$$

即 d_T 可行.

为证下降性,需证 $\nabla T(x, \bar{x})^T d_T < 0$. 注意到 $\nabla T(x, \bar{x})$ 的表达式及 $\bar{A}(x - \bar{x}) = 0$,有

$$\nabla T(x, \bar{x})^T d_T$$
$$= -\nabla T(x, \bar{x})^T \left[P\nabla f(x) - \frac{\alpha(f(x) - f(\bar{x}) + r)}{\| x - \bar{x} \|^2}(x - \bar{x}) \right]$$

$$=-\nabla T(x,\bar{x})^{\mathrm{T}}\Big\{P\nabla f(x)-\frac{\alpha(f(x)-f(\bar{x})+r)}{\|x-\bar{x}\|^2}\times$$

$$[(x-\bar{x})-\bar{A}^{\mathrm{T}}(\bar{A}\bar{A}^{\mathrm{T}})^{-1}\bar{A}(x-\bar{x})]\Big\}$$

$$=-\nabla T(x,\bar{x})^{\mathrm{T}}P\Big[\nabla f(x)-\frac{\alpha(f(x)-f(\bar{x})+r)}{\|x-\bar{x}\|^2}(x-\bar{x})\Big]$$

$$=-\|x-\bar{x}\|^{\alpha}\nabla T(x,\bar{x})^{\mathrm{T}}P\nabla T(x,\bar{x}).$$

由 $\mathrm{d}_T\neq 0$ 易知

$$P\nabla f(x)-\frac{\alpha(f(x)-f(\bar{x})+r)}{\|x-\bar{x}\|^2}(x-\bar{x})=P\nabla T(x,\bar{x})\neq 0,$$

从而由矩阵 P 的半正定性知，$\nabla T(x,\bar{x})^{\mathrm{T}}\mathrm{d}_T<0$.

定理 4.3.4 如果算法步骤 4 定义的 $\lambda_J\geqslant 0$，则 $(\bar{x}^*,\lambda)$ 是原问题 (4.1.1) 的 *KKT* 对. 其中 $\lambda=(\lambda_j:j=1,2,\cdots,m)$，当 $j\in J(\bar{x}^*)$ 时，λ_j 的取值为 (4.2.3) 式；当 $j\notin J(\bar{x}^*)$ 时，$\lambda_j=0$.

证明 由算法知，$\bar{x}^*=\bar{x}^k$ 是子问题 (4.2.1) 的一个局部极小点亦即 *KKT* 点，因此根据定理 4.3.1(2)，有 $\mathrm{d}(\bar{x}^*)=-P\nabla f(\bar{x}^*)=0$，即

$$\nabla f(\bar{x}^*)=\bar{A}^{\mathrm{T}}(\bar{A}\bar{A}^{\mathrm{T}})^{-1}\bar{A}\nabla f(\bar{x}^*)$$

成立.

注意到 $\bar{x}^*$ 满足约束，并由 λ_J 和 λ 定义，易知 $(\bar{x}^*,\lambda)$ 满足下列各式.

$$\nabla f(\bar{x}^*)+A^{\mathrm{T}}\lambda=0$$

$$A\bar{x}^*\leqslant b$$

$$\lambda\geqslant 0$$

$$\lambda^{\mathrm{T}}(A\bar{x}^*-b)=0.$$

即 $(\bar{x}^*,\lambda)$ 是问题 (4.1.1) 的 *KKT* 对.

在算法步骤 5 中，取

$$Q(x, \bar{x}^*, a_1^\mathrm{T}x - b_1, \cdots, a_m^\mathrm{T}x - b_m)$$

$$= \frac{1}{\parallel x - \bar{x}^* \parallel^\alpha}[f(x) - f(\bar{x}^*) + r + \alpha^3 \max_{1 \leqslant j \leqslant m}\{0, a_j^\mathrm{T}x - b_j\}]$$

$$(4.3.7)$$

其中, $r > 0$ 适当小; $\alpha \geqslant 1$ 适当大.

在点 x 处, 设

$$V = V(x) = \{j \mid a_j^\mathrm{T}x > b_j, j \in J\}.$$

又设

$$\Phi = \Phi(x) = \{j_0 \mid a_{j_0}^\mathrm{T}x - b_{j_0} = \max_{j \in V}\{a_j^\mathrm{T}x - b_j\}\}.$$

则 $\Phi \neq \varnothing$ 时, (4.3.7)式成为

$$Q(x, \bar{x}^*, a_1^\mathrm{T}x - b_1, \cdots, a_m^\mathrm{T}x - b_m) = \frac{1}{\parallel x - \bar{x}^* \parallel^\alpha}$$

$$\left[f(x) - f(\bar{x}^*) + r + \alpha^3 \sum_{j_0 \in \Phi}\lambda_{j_0}(a_{j_0}^\mathrm{T}x - b_{j_0}) \right] \quad (4.3.8)$$

其中

$$\lambda_{j_0} \geqslant 0, \qquad \sum_{j_0 \in \Phi}\lambda_{j_0} = 1.$$

下面讨论 $Q(x, \bar{x}^*, a_1^\mathrm{T}x - b_1, \cdots, a_m^\mathrm{T}x - b_m)$ 的性质. 为方便, 记 $Q(x, \bar{x}^*, a_1^\mathrm{T}x - b_1, \cdots, a_m^\mathrm{T}x - b_m)$ 为 $Q(x, \bar{x}^*, a_j)$.

又记集合 $E(0, 1) = \{x \mid \parallel x \parallel \leqslant 1, x \in R^n\}$, $\delta > 0$ 充分小.

定理 4.3.5 如果 $x \notin X \bigcup \delta E(0, 1)$, 但 $x \in \Omega$, 则当 α 充分大后, $\nabla_x Q(x, \bar{x}^*, a_j) \neq 0$.

证明 若存在 $x \notin X$, 则 $V \neq \varnothing$.

设 $x \notin X$, 由(4.3.8)式

$$\nabla_x Q(x, \bar{x}^*, a_j) = \frac{1}{\parallel x - \bar{x}^* \parallel^{\alpha+2}}\left[\parallel x - \bar{x}^* \parallel^2 \nabla f(x) + \right.$$

$$\alpha^3 \parallel x - \bar{x}^* \parallel^2 \sum_{j_0 \in \Phi} \lambda_{j_0} a_{j_0} - \alpha \left[f(x) - \right.$$

$$\left. f(\bar{x}^*) + r + \alpha^3 \sum_{j_0 \in \Phi} \lambda_{j_0} (a_{j_0}^T x - b_{j_0}) \right] (x - \bar{x}^*) \right],$$

所以,

$$(x - \bar{x}^*)^T \nabla_x Q(x, \bar{x}^*, a_j)$$

$$= \frac{1}{\parallel x - \bar{x}^* \parallel^\alpha} \left[(x - \bar{x}^*)^T \nabla f(x) + \alpha^3 \sum_{j_0 \in \Phi} \lambda_{j_0} a_{j_0}^T (x - \bar{x}^*) - \right.$$

$$\alpha \left[f(x) - f(\bar{x}^*) + r + \alpha^3 \sum_{j_0 \in \Phi} \lambda_{j_0} (a_{j_0}^T x - b_{j_0}) \right] \right]$$

$$= \frac{1}{\parallel x - \bar{x}^* \parallel^\alpha} \left[(x - \bar{x}^*)^T \nabla f(x) - \alpha (f(x) - f(\bar{x}^*) + r) + \right.$$

$$\alpha^3 \sum_{j_0 \in \Phi} \lambda_{j_0} a_{j_0}^T (x - \bar{x}^*) - \alpha^4 \sum_{j_0 \in \Phi} \lambda_{j_0} (a_{j_0}^T x - b_{j_0}) \right].$$

由 $x \notin X \bigcup \delta E(0, 1)$, 而 $\bar{x}^* \in X$ 知 $a_{j_0}^T (x - \bar{x}^*) > 0$, $(a_{j_0}^T x - b_{j_0}) \geqslant \delta$, 所以 α 充分大后有

$$\alpha^3 \sum_{j_0 \in \Phi} \lambda_{j_0} a_{j_0}^T (x - \bar{x}^*) - \alpha^4 \sum_{j_0 \in \Phi} \lambda_{j_0} (a_{j_0}^T x - b_{j_0}) < 0$$

又由假设 4.1.1 和 4.1.2 知, $(x - \bar{x}^*)^T \nabla f(x)$ 和 $f(x) - f(\bar{x}^*) + r$ 有界. 因此当 α 充分大后有

$$(x - \bar{x}^*)^T \nabla_x Q(x, \bar{x}^*, a_j) < 0,$$

即

$$\nabla_x Q(x, \bar{x}^*, a_j) \neq 0, \quad \forall x \notin X.$$

定理 4.3.6 设 $x \in X \bigcap S_1(\bar{x}^*)$, 则对适当大的 α, 有 $\nabla_x Q(x, \bar{x}^*, a_j) \neq 0$.

证明 因为 $x \in X$, 所以

$$Q(x, \bar{x}^*, a_j) = \frac{1}{\| x - \bar{x}^* \|^{\alpha}} [f(x) - f(\bar{x}^*) + r].$$

又因为 $x \in S_1(\bar{x}^*)$，即 $f(x) \geqslant f(\bar{x}^*)$，$x \neq \bar{x}^*$，所以

$$(x - \bar{x}^*)^T \nabla_x Q(x, \bar{x}^*, a_j)$$

$$= \frac{1}{\| x - \bar{x}^* \|^{\alpha}} [(x - \bar{x}^*)^T \nabla f(x) - \alpha (f(x) - f(\bar{x}^*) + r)]$$

$$\leqslant \frac{1}{\| x - \bar{x}^* \|^{\alpha}} [(x - \bar{x}^*)^T \nabla f(x) - \alpha r],$$

根据假设 4.1.1 和 4.1.2 知，$| (x - \bar{x}^*)^T \nabla f(x) |$ 有界，设

$$\overline{D} = \max_{x \in X} \| x - \bar{x}^* \|, \quad \| \nabla f(x) \| \leqslant K, \quad x \in X \quad (4.3.9)$$

则当 $\alpha > \dfrac{K \overline{D}}{r}$ 时，必有 $(x - \bar{x}^*)^T \nabla_x Q(x, \bar{x}^*, a_j) < 0$ 成立，亦即 $\nabla_x Q(x, \bar{x}^*, a_j) \neq 0$ 成立.

定理 4.3.7 设 $\bar{x}^* \in X$，是 $f(x)$ 的局部极小点，$x_1, x_2 \in X \cap S_1(\bar{x}^*)$ 且 $\| x_1 - \bar{x}^* \| \geqslant \| x_2 - \bar{x}^* \| + \varepsilon$，这里 $\varepsilon > 0$. 则当

$$\alpha > \max \left\{ 1, \frac{\ln r - \ln (K \overline{D} + r)}{\ln (\overline{D} - \varepsilon) - \ln (\overline{D})} \right\} \quad (4.3.10)$$

时有

$$Q(x_1, \bar{x}^*, a_j) < Q(x_2, \bar{x}^*, a_j)$$

证明 因为 $x_1, x_2 \in X \cap S_1(\bar{x}^*)$，所以 $f(x_1) - f(\bar{x}^*) + r > 0$，$f(x_2) - f(\bar{x}^*) + r > 0$，且

$$Q(x_i, \bar{x}^*, a_j) = \frac{1}{\| x_i - \bar{x}^* \|^{\alpha}} [f(x_i) - f(\bar{x}^*) + r], \quad i = 1, 2.$$

当 $f(x_1) \leqslant f(x_2)$ 时，由于 $\| x_1 - \bar{x}^* \| > \| x_2 - \bar{x}^* \|$，显然

对 $\alpha \geqslant 1$ 有

$$\frac{f(x_2) - f(\bar{x}^*) + r}{f(x_1) - f(\bar{x}^*) + r} \geqslant 1 > \left[\frac{\| x_2 - \bar{x}^* \|}{\| x_1 - \bar{x}^* \|} \right]^{\alpha},$$

即 $Q(x_1, \bar{x}^*, a_j) < Q(x_2, \bar{x}^*, a_j)$ 成立.

当 $f(x_1) > f(x_2)$ 时, 由 $\| x_1 - \bar{x}^* \| \geqslant \| x_2 - \bar{x}^* \| + \varepsilon$, 可得

$$\frac{\| x_2 - \bar{x}^* \|}{\| x_1 - \bar{x}^* \|} \leqslant 1 - \frac{\varepsilon}{D} \tag{4.3.11}$$

而

$$\frac{f(x_2) - f(\bar{x}^*) + r}{f(x_1) - f(\bar{x}^*) + r} \geqslant \frac{r}{K\overline{D} + r} \tag{4.3.12}$$

故当 $\alpha > \dfrac{\ln r - \ln(K\overline{D} + r)}{\ln(\overline{D} - \varepsilon) - \ln(\overline{D})}$ 有

$$\frac{r}{K\overline{D} + r} > \left(1 - \frac{\varepsilon}{D} \right)^{\alpha},$$

由(4.3.11)和(4.3.12)式得到

$$\frac{f(x_2) - f(\bar{x}^*) + r}{f(x_1) - f(\bar{x}^*) + r} > \left[\frac{\| x_2 - \bar{x}^* \|}{\| x_1 - \bar{x}^* \|} \right]^{\alpha},$$

从而, $Q(x_1, \bar{x}^*, a_j) < Q(x_2, \bar{x}^*, a_j)$.

定理 4.3.7 说明在可行域 $X \cap S_1(\bar{x}^*)$ 上, 只要 α 的取值适当大, 则离当前局部极小点 $\bar{x}^*$ 越远的地方, $Q(x, \bar{x}^*, a_j)$ 的函数值越小.

定理 4.3.8 如果 $\bar{x}^* \in L(P)$ 但是 $\bar{x}^* \notin G(P)$, 则必有合适的 $r > 0$ 和 $\alpha \geqslant 1$ 使得存在 $x' \in X \cap S_2(\bar{x}^*)$, 成为 $Q(x, \bar{x}^*, a_j)$ 的极小点.

证明 因为 $\bar{x}^* \notin G(P)$, 所以 $X \cap S_2(\bar{x}^*) \neq \varnothing$.

设 x_G 是问题(4.1.1)的全局解, 即 $x_G \in G(P)$, 则有 $f(x_G) < f(\bar{x}^*)$. 令 $r > 0$, 取

$$0 < r < f(\bar{x}^*) - f(x_G) \tag{4.3.13}$$

由于 $x_G \in X$，并由(4.3.13)式知

$$Q(x_G, \bar{x}^*, a_j) = \frac{1}{\| x_G - \bar{x}^* \|}[f(x_G) - f(\bar{x}^*) + r] < 0.$$

记点 $\bar{x}^*$ 的邻域为 $N(\bar{x}^*)$. 由于 $Q(x, \bar{x}^*, a_j)$ 在闭区域 $\Omega \backslash N(\bar{x}^*)$ 上连续，$Q(x, \bar{x}^*, a_j)$ 必在 $\Omega \backslash N(\bar{x}^*)$ 上存在极小点 x'，且

$$Q(x', \bar{x}^*, a_j) \leqslant Q(x_G, \bar{x}^*, a_j) < 0 \qquad (4.3.14)$$

下证 $x' \in X \bigcap S_2(\bar{x}^*)$. 分几种情况讨论.

如果 $x \notin X$，则 $\Phi \neq \varnothing$，由(4.3.8)式，

$$Q(x, \bar{x}^*, a_j) = \frac{1}{\| x - \bar{x}^* \|^\alpha}\Big[f(x) - f(\bar{x}^*) + r + \alpha^3 \sum_{j_0 \in \Phi}(a_{j_0}^{\mathrm{T}} x - b_{j_0})\Big],$$

因为 $a_{j_0}^{\mathrm{T}} x - b_{j_0} > 0$ $(j_0 \in \Phi)$，故当 α 充分大后，一定有 $Q(x, \bar{x}^*, a_j) > 0$，此与(4.3.14)式矛盾. 所以 x' 不会在可行域的外面.

当 $x = \bar{x}^*$ 时，由于 $\lim\limits_{x \to \bar{x}^*} Q(x, \bar{x}^*, a_j) = +\infty$，所以 $x' \notin N(\bar{x}^*)$.

当 $x \in X \bigcap S_1(\bar{x}^*)$ 时，因为 $f(x) \geqslant f(\bar{x}^*)$ 所以

$$Q(x, \bar{x}^*, a_j) = \frac{1}{\| x - \bar{x}^* \|^\alpha}[f(x) - f(\bar{x}^*) + r] > 0.$$

与(4.3.14)式矛盾. 故而 $x' \notin X \bigcap S_1(\bar{x}^*)$.

综上所述，结合 x' 的存在性知 $x' \in X \bigcap S_2(\bar{x}^*)$.

§4.4 非线性约束全局优化问题及解法

考虑一般形式的带不等式约束的非线性规划全局最优问题：

$$\begin{aligned} &\min f(x) \\ &s.t. \ g_j(x) \leqslant 0, \ j = 1, 2, \cdots, m \end{aligned} \qquad (4.4.1)$$

这里 $x \in R^n$；$f(x)$ 和 $g_j(x)(j=1, 2, \cdots, m)$：$R^n \to R$. J 和 $J(x)$ 的意义同线性约束的情况.

记可行域 $S = \{x \mid g_j(x) \leqslant 0, j \in J\}$. int$S$ 表示可行域 S 的内点集合. ∂S 表示 S 的边界.

$L(Np)$ 为问题(4.4.1)的 KKT 点集合.

$G(NP)$ 为问题(4.4.1)的全局最优解集合.

对目标函数 $f(x)$，假设 4.1.1 和假设 4.1.2 仍然成立. 因而可假设存在有界闭域 Ω 包含可行域 S.

假设 4.1.3 换为：

假设 4.4.1 在 x 处，有效约束的梯度线性无关，即 $(\nabla g_j(x)$：$j \in J(x))$ 线性无关.

增加下面的假设：

假设 4.4.2 函数 $g_j(x)(j=1, 2, \cdots, m)$ 在 Ω 上一阶连续可微.

假设 4.4.3 $G(NP) \subset$ int$S$，且 $G(NP)$ 的个数有限.

设 $x^* \in L(NP)\backslash G(NP)$；$x_G \in G(NP)$，则 $f(x^*) - f(x_G) > 0$.

任取 $x_G \in G(NP)$，令

$$\mathrm{d}(x_G) = \min_{x \in \partial S} \| x_G - x \|, \quad d = \min_{x_G \in G(NP)} \mathrm{d}(x_G).$$

因为 x_G 是 S 的内点，所以 $d > 0$.

取参数

$$0 < r < \min\{d, f(x^*) - f(x_G)\}, \quad \alpha \geqslant 1 \quad (4.4.2)$$

现在给出算法如下：

算法：

步骤 1 取 $\varepsilon_0 > 0$，充分小. 在可行域内任取初始点 x_0.

步骤 2 以 x_0 为起点，用下降算法求出约束问题(4.4.1)的一个 KKT 点 x^*.

步骤 3 在点 x^* 处构造变换函数

$$U(x, x^*, g(x)) = \frac{1}{\| x - x^* \|^\alpha}\left[f(x) - f(x^*) + r + \right.$$

$$\alpha^3 \max_{1 \leqslant j \leqslant m} \{0,\, g_j(x)+r\}] \qquad (4.4.3)$$

其中参数 r 和 α 的取值满足 $(4.4.2)$ 式. 并对参数 r, α 适当取值.

步骤 4 如果 $r < \varepsilon_0$,则停,x^* 可以认为是所求全局解. 否则求解
问题

$$\min_{x \in \Omega} U(x,\, x^*,\, g(x)) \qquad (4.4.4)$$

步骤 5 如果问题 $(4.4.4)$ 有解 x',则令 $x_0 = x'$,转步骤 2. 否则
令 $r = \dfrac{1}{2}r$,$\alpha = 10\alpha$,转步骤 4.

在表达式 $(4.4.5)$ 中,若不止一个 $g_l(x)$ 等于 $\max\limits_{1 \leqslant j \leqslant m} g_j(x)$ 时,定
义集合

$$Vg = \{l \mid g_l(x) = \max_{1 \leqslant j \leqslant m} g_j(x)\} > 0,$$

则 $Vg \neq \varnothing$ 时

$$U(x,\, x^*,\, g(x)) = \frac{1}{\| x - x^* \|^{\alpha}}[f(x) - f(x^*) + r +$$

$$\alpha^3 \sum_{l \in Vg} (\, g_l(x) + r)] \qquad (4.4.5)$$

下面讨论算法及变换函数 $U(x,\, x^*,\, g(x))$ 的性质.

定理 4.4.1 如果 $x \notin S$,但 $x \in \Omega$,则当 α 充分大后,$\nabla_x U(x,\, x^*,\, g(x)) \neq 0$.

证明 因为 $x \notin S$,所以必有 $Vg \neq \varnothing$.
由 $U(x,\, x^*,\, g(x))$ 的定义,有

$$\nabla_x U(x,\, x^*,\, g(x)) = \frac{1}{\| x - x^* \|^{\alpha+2}}\Big\{\Big[\nabla f(x) + \alpha^3 \sum_{l \in Vg} \nabla g_l(x)\Big]$$

$$\| x - x^* \|^2 - \alpha\Big[f(x) - f(x^*) + r +$$

$$\alpha^3 \sum_{l \in Vg} (g_l(x) + r)\Big](x - x^*)\Big\},$$

进而

$$(x-x^*)^T \nabla_x U(x, x^*, g(x))$$

$$= \frac{1}{\|x-x^*\|^\alpha} \Big\{ (x-x^*)^T \nabla f(x) + \alpha^3 (x-x^*)^T \sum_{l \in V_g} \nabla g_l(x) - $$

$$\alpha \Big[f(x) - f(x^*) + r + \alpha^3 \sum_{l \in V_g} (g_l(x) + r) \Big] \Big\}$$

$$= \frac{1}{\|x-x^*\|^\alpha} \Big[(x-x^*)^T \nabla f(x) - \alpha (f(x) - f(x^*) + r) + $$

$$\alpha^3 (x-x^*)^T \sum_{l \in V_g} \nabla g_l(x) - \alpha^4 \sum_{l \in V_g} (g_l(x) + r) \Big]$$

$$\leqslant \frac{1}{\|x-x^*\|^\alpha} \Big[(x-x^*)^T \nabla f(x) - \alpha (f(x) - f(x^*) + r) + $$

$$\alpha^3 (x-x^*)^T \sum_{l \in V_g} \nabla g_l(x) - \alpha^4 \sum_{l \in V_g} r \Big].$$

由假设条件知 $(x-x^*)^T \nabla f(x)$，$f(x) - f(x^*) + r$ 和 $(x-x^*)^T \nabla g_l(x)$ 都有界，且 $r > 0$。通过比较 α 的指数可知，当 α 充分大后，必有

$$(x-x^*)^T \nabla_x U(x, x^*, g(x)) < 0.$$

意即，α 充分大后 $\nabla_x U(x, x^*, g(x)) \neq 0$。

定理 4.1.1 说明，在一定条件下子问题(4.4.4)如果有解，一定在可行域 S 内取得。

定理 4.4.2 设 $x \in S$ 且 $f(x) \geqslant f(x^*)$，$x \neq x^*$，即 $x \in S \bigcap S_1$，则当 α 充分大，$r > 0$ 适当小，有 $\nabla_x U(x, x^*, g(x)) \neq 0$。

证明 因为 $x \in S$，所以存在 $r > 0$ 充分小，使得

$$U(x, x^*, g(x)) = \begin{cases} \dfrac{1}{\|x-x^*\|^\alpha} [f(x) - f(x^*) + r], & x \in \text{int}S \\[3mm] \dfrac{1}{\|x-x^*\|^\alpha} [f(x) - f(x^*) + r + \alpha^3 r], & x \in \partial S \end{cases}$$

$$(4.4.6)$$

无论 $U(x, x^*, g(x))$ 的形式如何，总有

$$\nabla_x U(x, x^*, g(x)) = \frac{1}{\parallel x-x^* \parallel^{\alpha+2}} \{\nabla f(x) \parallel x-x^* \parallel^2 -$$

$$\alpha [f(x)-f(x^*)+r](x-x^*)\},$$

等式两边与 $x-x^*$ 作内积，并注意到 $f(x) > f(x^*)$，得

$$(x-x^*)^{\mathrm{T}} \nabla_x U(x, x^*, g(x)) = \frac{1}{\parallel x-x^* \parallel^{\alpha}}$$

$$[(x-x^*)^{\mathrm{T}} \nabla f(x) - \alpha (f(x)-f(x^*)+r)]$$

$$\leqslant \frac{1}{\parallel x-x^* \parallel^{\alpha}} [(x-x^*)^{\mathrm{T}} \nabla f(x) - \alpha r].$$

根据假设条件，$(x-x^*)^{\mathrm{T}} \nabla f(x)$ 有界，$r > 0$，因而当 α 充分大必有

$$(x-x^*)^{\mathrm{T}} \nabla_x U(x, x^*, g(x)) < 0$$

即 $\nabla_x U(x, x^*, g(x)) \neq 0$.

定理 4.4.2 说明在一定的条件下，子问题(4.4.4)在区域 $S \bigcap S_1$ 没有解.

定理 4.4.3 如果 $x^* \notin G(NP)$，则必有合适的 $r > 0$ 和 $\alpha \geqslant 1$ 使得函数 $U(x, x^*, g(x))$ 在区域 $S \bigcap S_2(x^*)$ 上有极小点 x'.

证明 因为 $x^* \notin G(NP)$，所以 $S \bigcap S_2(x^*) \neq \varnothing$，特别地 $x_G \in S \bigcap S_2(x^*)$.

记点 x^* 的邻域为 $\delta(x^*)$，这是一个开区域. 由 $U(x, x^*, g(x))$ 的定义和 $f(x), g_j(x)(j = 1, 2, \cdots, m)$ 的连续可微性知，$U(x, x^*, g(x))$ 在有界闭域 $\Omega \backslash \delta(x^*)$ 连续，所以一定有 $x' \in \Omega \backslash \delta(x^*)$ 使得 $f(x') = \min\limits_{x \in \Omega \backslash \delta(x^*)} f(x)$.

另一方面取 $x_G \in G(NP)$. 由于 $G(NP) \subset intS$，所以由 r 的定义 (4.4.2) 式及 $U(x, x^*, g(x))$ 的定义得

$$U(x_G,\ x^*,\ g(x)) = \frac{1}{\parallel x_G - x^* \parallel^{\alpha}}[f(x_G) - f(x^*) + r] < 0.$$

从而一定有

$$U(x',\ x^*,\ g(x)) \leqslant U(x_G,\ x^*,\ g(x)) < 0 \qquad (4.4.7)$$

根据 x' 的存在性, 下面用排除法证明 $x' \in S \cap S_2(x^*)$.

当 $x \in \Omega\backslash S$, 即 $x \notin S$ 时,

$$U(x,\ x^*,\ g(x)) = \frac{1}{\parallel x - x^* \parallel^{\alpha}}[f(x) - f(x^*) +$$

$$r + \alpha^3(g_l(x) + r)],$$

因为 $g_l(x) > 0, r > 0$, 所以 α 适当大时必有 $f(x) - f(x^*) + r + \alpha^3(g_l(x) + r) > 0$, 亦即 $U(x,\ x^*,\ g(x)) > 0$, 此与 (4.4.7) 式矛盾. 故 $x' \notin \Omega\backslash S$.

而 $x \in S \cap S_1(x^*)$ 时, 存在适当小的 $r > 0$ 和 $\alpha \geqslant 1$ 根据 (4.4.6) 式得 $U(x,\ x^*,\ g(x)) \geqslant r > 0$, 此不等式亦与 (4.4.7) 式相矛盾. 所以 $x' \notin S \cap S_1(x^*)$.

显然 $\Omega\backslash\delta(x^*) \subseteq (\Omega\backslash S) \cup (S \cap S_1(x^*)) \cup (S \cap S_2(x^*))$. 但是 $S \cap S_2(x^*) \subset \Omega\backslash\delta(x^*)$. 故而由 $x' \in \Omega\backslash\delta(x^*)$, 却又有 $x' \notin \Omega\backslash S$ 以及 $x' \notin S \cap S_1(x^*)$ 知, 必定有 $x' \in S \cap S_2(x^*)$.

由上面的定理我们有推论

推论 4.4.1 在定理 4.4.3 的条件下, 存在 $x' \in S$ 使得 $f(x') < f(x^*)$.

根据算法, 以 x' 为新的初始点, 求解问题(4.4.1), 可以得到更好的 KKT 点.

定理 4.4.4 若 $x^* \notin G(NP)$, 则存在适当的 r 和 α 的值, 使得通过算法可以找到 $x_1^* \in L(NP)$ 并且 有 $f(x_1^*) < f(x^*)$.

第五章 求约束极值问题的修正共轭梯度投影算法

§5.1 问题的介绍

本章考虑带约束的非线性优化问题

$$\min \{ f(x) : g_j(x) \leqslant 0, \, j \in I, \, x \in R^n \} \qquad (5.1.1)$$

这里 $f(x)$, $g_j(x) : R^n \to R$, $j \in I = \{1, 2, \cdots, m\}$.

我们知道拟牛顿方法 [49, 26, 55] 由于它的超线性收敛的特性而成为求解无约束最优化问题的有效方法之一. 目前对这种方法在理论及计算上的各种改进仍然是比较活跃的. 拟牛顿法与其他方法的结合用于求解问题 (5.1.1) 是拟牛顿方法改进研究的重要方向之一. 一般称为约束拟牛顿法. 共轭投影梯度法与拟牛顿法的结合是其中一个方法. 然而, 约束拟牛顿方法中为了证明局部超线性收敛速度需要一个很强的假设条件——严格互补条件, 这是一个很难验证的条件. Bonnans 和 Launay [7] 提出了一个不用严格互补条件而且超线性收敛的方法, 但是这个方法每迭代一步需要求解两个二次子规划. 一般地讲, 在同一步迭代中约束拟牛顿法的搜索方向是在两个不同的方法中作出选择, 即, 在某些条件下是改进的 (或称修正的) 拟牛顿方向, 在另一些条件下是梯度的投影类方向, 参见 [77, 83]. 这样由于需要计算两次投影矩阵而导致了计算量的增加. 还有一些方法如序列二次规划 (SQP)([6, 49, 26, 55) 是一类求解问题 (5.1.1) 较有效的方法, 但为了克服 Maratos 效应 [48], 算法往往需要进行高阶修正, 而这又导致了计算量的增加.

　　为了克服上述的两个缺陷——强假设条件和大计算量,本章给出一个修正共轭投影梯度法,其特点是不用严格互补条件而证明了算法的超线性收敛性,尤其是每步迭代只需计算一次共轭投影矩阵避免了求解二次规划或求两个投影矩阵,从而减少了计算量. 即本算法在理论上和计算量上均有所改进. 此方法用了 ε 积极约束集技巧并且给出了显式搜索方向,算法是和谐有效的. 在一般的假设条件下证明了算法的收敛性和超线性收敛性.

　　本章的结构如下:第二节给出了算法;第三节讨论了算法的收敛性;第四节分析算法的收敛速度;最后第五节给出了三个算例的数值结果.

§5.2　算法

　　记 $X = \{x: g_j(x) \leqslant 0\}$, $J(x) = \{j \in I: g_j(x) \geqslant -\varepsilon\}$. 本章需要下列两个假设.

　　假设 5.2.1　对任意 $j \in I$, $f(x)$ 和 $g_j(x)$ 连续可微;

　　假设 5.2.2　对 $x \in X$, $\{\nabla g_j(x), j \in J(x)\}$ 是线性无关的.

　　首先介绍一些记号. 在当前点 x_k,我们定义

$$A_k = (\nabla g_j(x_k), j \in J(x_k)),$$

$$g_{J_k} = (g_j(x_k), j \in J(x_k))^{\mathrm{T}}.$$

　　关于给定的对称正定矩阵 H_k 的共轭投影矩阵为

$$P_k = H_k - H_k A_k B_k \qquad (5.2.1)$$

其中

$$B_k = (A_k^{\mathrm{T}} H_k A_k)^{-1} A_k^{\mathrm{T}} H_k \qquad (5.2.2)$$

我们定义

$$d_k^0 = -P_k \nabla f(x_k) - B_k^{\mathrm{T}} g_{J_k} \qquad (5.2.3)$$

和

$$\lambda_k = -B_k \nabla f(x_k) + (A_k^{\mathrm{T}} H_k A_k)^{-1} g_{J_k} = \lambda_k^1 + \lambda_k^2 \quad (5.2.4)$$

这里，$\lambda_k^1 = -B_k \nabla f(x_k)$，$\lambda_k^2 = (A_k^{\mathrm{T}} H_k A_k)^{-1} g_{J_k}$。

如果集合 $J(x_k) = \varnothing$，算法就是通常的拟牛顿方法。下面，我们总是假设 $J(x_k) \neq \varnothing$。

为简单起见，将 x_k，P_k，A_k，$\cdots$ 记为 x，P，A，$\cdots$；J 代表 $J(x)$。

引理 5.2.1 P 是一个半正定矩阵，并且 $PA = 0$，$BA = E$，其中 E 是一个 $|J| \times |J|$ 单位矩阵。

证明 由 P 和 B 的定义，

$$PA = HA - HABA = HA - AH(A^{\mathrm{T}} HA)^{-1} A^{\mathrm{T}} HA = 0；$$

$$BA = (A^{\mathrm{T}} HA)^{-1} A^{\mathrm{T}} HA = E。$$

定理 5.2.1 如果 $x \in X$，$d^0 = 0$，$\lambda \geqslant 0$，则 x 是问题 (5.1.1) 的一个 KKT 点。

证明 根据 $d^0 = 0$，有

$$0 = -H \nabla f(x) + HA(A^{\mathrm{T}} HA)^{-1} A^{\mathrm{T}} H \nabla f(x) -$$
$$HA(A^{\mathrm{T}} HA)^{-1} g_J = -H \nabla f(x) - HA\lambda$$

和

$$0 = A^{\mathrm{T}} d^0 = -A^{\mathrm{T}} P \nabla f(x) - (BA)^{\mathrm{T}} g_J = -g_J。$$

因此，存在 $\lambda \geqslant 0$ 使得 $\nabla f(x) + A\lambda = 0$ 和 $\lambda_j g_j = 0$，$j \in J$。证毕。

现在给出算法如下：

Step 0 给定 $\alpha \in \left(0, \dfrac{1}{2}\right)$，$\beta$，$\eta \in (0, 1)$，$\theta \in \left(\dfrac{1}{2}, 1\right)$；$\varepsilon$，$\delta > 0$。选择 $x_0 \in X$ 和一个对称正定矩阵 H_0，令 $k = 0$。

Step 1.1 令 $i = 0$，$\varepsilon^{ki} = \varepsilon^k$。

Step 1.2 如果

$$\det(A_{ki}^{\mathrm{T}}A_{ki})>\varepsilon^{ki},$$

其中 $A_{ki}=(\nabla g_j(x_k)：j\in J_{ki})$ 及 $J_{ki}=\{j\in I：g_j(x_k)>-\varepsilon^{ki}\}$，则令 $A_k=A_{ki}$，$J_k=J_{ki}$，$\varepsilon^k=\varepsilon^{ki}$，转到 Step 2. 否则

Step 1.3 令 $i=i+1$，$\varepsilon^{ki}=\dfrac{1}{2}\varepsilon^{k(i-1)}$，转到 Step 1.2.

Step 2 根据(5.2.3)和(5.2.4)计算 d_k^0，λ_k. 如果 $d_k^0=0$，$\lambda_k\geqslant 0$，则停.

Step 3 如果 $\|d_k^0\|\leqslant\delta$，$\lambda_{kj}\geqslant-\eta\|d_k^0\|$ 并且

$$\begin{cases} f(x_k+d^0)\leqslant f(x_k)+\alpha\,(\nabla f(x_k)^{\mathrm{T}}d_k^0-\lambda_k^{\mathrm{T}}g_{J_k}) \\ g_j(x_k+d^0)\leqslant 0,\ j\in I \end{cases} \tag{5.2.5}$$

则令 $x_{k+1}=x_k+d_k^0$ and $\delta=\dfrac{1}{2}\delta$. 转到 Step 7.

Step 4 令 $U_k=(u_{kj},\ j\in J_k)^{\mathrm{T}}$，其中

$$u_{kj}=\begin{cases} \lambda_{kj}^1,\ \lambda_{kj}^1<0, \\ -|g_j(x_k)|,\ \lambda_{kj}^1\geqslant 0. \end{cases}$$

令 $d_k^1=-P_k\nabla f(x_k)+B_k^{\mathrm{T}}U_k$ 及 $d_k^2=-P_k\nabla f(x_k)-B_k^{\mathrm{T}}\|d^1\|e$，这里 $e=(1,1,\cdots,1)^{\mathrm{T}}$.

Step 5 设

$$d_k=(1-\rho_k)d_k^1+\rho_k d_k^2 \tag{5.2.6}$$

其中 $\rho_k=\max\{\rho\in(0,1]：\nabla f(x_k)^{\mathrm{T}}((1-\rho)d_k^1+\rho d_k^2)\leqslant\theta\,\nabla f(x_k)^{\mathrm{T}}d_k^1\}$.

Step 6 序列 $\{1,\beta,\beta^2,\cdots\}$ 中第一个满足下列关于 t 的不等式的项记为 t_k，

$$\begin{cases} f(x_k+t\,d_k)\leqslant f(x_k)+\alpha\,t\,\nabla f(x_k)^{\mathrm{T}}d_k^1 \\ g_j(x_k+td_k)\leqslant 0,\ j\in I \end{cases} \tag{5.2.7}$$

并且令 $x_{k+1} = x_k + t_k d_k$.

Step 7 根据拟牛顿公式修正 H_k 为 H_{k+1}. 令 $k = k+1$ 并返回 Step 1.

根据假设 5.2.2,不难得到下面的引理.

引理 5.2.2 存在正数 $\epsilon > 0$ 使得对所有 k,有 $\epsilon^k \geqslant \epsilon$.

§5.3 算法的收敛性

本节,讨论算法的全局收敛性. 我们需要增加另外的假设,并且此假设在本章的后半部分一直成立.

假设 5.3.1 对任意 k 和 $y \in R^n$,有 $a\|y\|^2 \leqslant y^T H_k y \leqslant b\|y\|^2$ 成立,其中常数 $b \geqslant a > 0$;

假设 5.3.2 算法产生的点列 $\{x_k\}$ 有界.

引理 5.3.1 序列 $\{P_k\}$,$\{d_k^0\}$,$\{d_k\}$ 及 $\{\lambda_k\}$ 都是有界的.

证明 由假设 5.3.1 知,$\|H_k\|$ 是有界的. 所以根据(5.2.1),(5.2.2)式,要证 P_k 有界,只需证明行列式 $\det((A_k^T H_k A_k)^{-1})$ 是有界的. 用反证法. 如果此行列式无界,则存在一个子集 K 使得

$$x_k \to x^* \in X, \qquad \det((A_k^T H_k A_k)^{-1}) \to \infty, \qquad k \xrightarrow{K} \infty.$$

这说明

$$\det(A_k^T H_k A_k) \to 0, \qquad k \xrightarrow{K} \infty,$$

亦即 $\det(A_k^T A_k) \to 0$,$(k \xrightarrow{K} \infty)$. 这和假设 2 矛盾. 根据假设 1 和 $\{P_k\}$ 的有界性,易知其余的结论是正确的.

引理 5.3.2 $d_k^0 = 0$,$\lambda_k \geqslant 0 \iff d_k^1 = 0$.

证明 首先令 $d_k^0 = 0$,$\lambda_k \geqslant 0$. 根据定理 5.2.1 的证明和 (5.2.3) 式,可得 $g_{J_k} = 0$,$P \nabla f(x_k) = 0$. 进一步,由 U_k 和 d_k^1 的定义,易知 $d_k^1 = 0$.

反之,由 $d_k^1 = 0$,我们可得

$$\nabla f(x_k)^{\mathrm{T}} d_k^1 = -\nabla f(x_k)^{\mathrm{T}} P_k \nabla f(x_k) + \nabla f(x_k)^{\mathrm{T}} B_k^{\mathrm{T}} U_k$$

$$= -\parallel P_k^{\frac{1}{2}} \nabla f(x_k) \parallel^2 - \sum_{j \in J_k} \lambda_{kj}^1 u_j$$

$$= -\parallel P_k^{\frac{1}{2}} \nabla f(x_k) \parallel^2 - \sum_{\lambda_{kj}^1 < 0} (\lambda_{kj}^1)^2 - \sum_{\lambda_{kj}^1 \geqslant 0} \lambda_{kj}^1 \mid g_j \mid$$

$$= 0.$$

因此, $P_k \nabla f(x_k) = 0$, $\lambda_k^1 \geqslant 0$. 根据算法 Step 4, 由 $d_k^1 = 0$ 和 $P_k \nabla f(x_k) = 0$ 可得 $B_k^{\mathrm{T}} U_k = 0$, 从而 $U_k = 0$, 又由 U_k 的定义知此时 $g_{J_k} = 0$. 所以有 $d_k^0 = 0$ 和 $\lambda_k \geqslant 0$ 成立.

引理 5.3.3 如果 x_k 不是问题(5.1.1)的 KKT 点,则

(i) $\nabla f(x_k)^{\mathrm{T}} d_k^0 - \lambda_k^{\mathrm{T}} g_{J_k} \leqslant 0, \nabla f(x_k)^{\mathrm{T}} d_k^1 < 0$;

(ii) $\nabla g_j(x_k)^{\mathrm{T}} d_k^1 \leqslant 0, \nabla g_j(x_k)^{\mathrm{T}} d_k^2 < 0, j \in J_k$.

证明 (i) 由于 (d_k^0, λ_k) 是下列二次规划的 KKT 对

$$(QP) \quad \min \frac{1}{2} d^{\mathrm{T}} H_k^{-1} d + \nabla f(x_k)^{\mathrm{T}} d$$

$$s.t. \quad g_j(x_k) + \nabla g_j(x_k)^{\mathrm{T}} d = 0, j \in J_k,$$

所以有

$$\nabla f(x_k)^{\mathrm{T}} d_k^0 - \lambda_k^{\mathrm{T}} g_{J_k} = \nabla f(x_k)^{\mathrm{T}} d_k^0 + \lambda_k^{\mathrm{T}} A_{J_k} d_k^0$$

$$= -(d_k^0)^{\mathrm{T}} H_k^{-1} d_k^0 \leqslant 0.$$

现在我们证明 $\nabla f(x_k)^{\mathrm{T}} d_k^1 < 0$. 因为 x_k 不是 KKT 点,则或者有 $d_k^0 \neq 0$ 或者存在 $j \in J_k$ 使得 $\lambda_{kj}^1 < 0$. 因此有

$$\nabla f(x_k)^{\mathrm{T}} d_k^1 = \nabla f(x_k)^{\mathrm{T}} (d_k^0 + B^{\mathrm{T}} g_{J_k} + B^{\mathrm{T}} U_k)$$

$$= -(d_k^0)^{\mathrm{T}} H_k^{-1} d_k^0 - \sum_{\lambda_{kj}^1 < 0} (\lambda_{kj}^1)^2 - \sum_{\lambda_{kj}^1 \geqslant 0} \lambda_{kj}^1 \mid g_j \mid$$

$$< 0.$$

(ii) 既然 x_k 不是 KKT 点,那么有 $d_k^1 \neq 0$. 因而,

$$A_{J_k}^T d_k^1 = U_k \leqslant 0, \qquad A_{J_k}^T d_k^2 = -\|d_k^1\|e < 0 \qquad (5.3.1)$$

根据(5.2.6)式和引理 5.3.3,显然有下面的引理成立.

引理 5.3.4 如果 x_k 不是问题(5.1.1)的 KKT 点,则有

(i) $\nabla f(x_k)^T d_k < 0$.

(ii) $\nabla g_j(x_k)^T d_k < 0$, $j \in J(x_k)$.

对于算法中的 Step 6 有如下引理.

引理 5.3.5 如果 x_k 不是问题(5.1.1)的 KKT 点,则存在 $\bar{t} > 0$ 使得对任意 $t \in [0, \bar{t}]$,有(5.2.7)式成立.

证明 根据微分中值定理有,

$$f(x_k + td_k) = f(x_k) + t\nabla f(x_k + \xi d_k)^T d_k, \quad \xi \in [0, t].$$

由于 $f(x)$ 是连续可微的,并且有 $\nabla f(x_k)^T d_k \leqslant \theta \nabla f(x_k) d_k^1 < 0$ 及 $\alpha/\theta < 1$,所以存在 $t_0 > 0$ 使得

$$f(x_k + td_k) \leqslant f(x_k) + t\frac{\alpha}{\theta}\nabla f(x_k)^T d_k$$

$$\leqslant f(x_k) + \alpha t \nabla f(x_k)^T d_k^1. \qquad \forall t \in [0, t_0].$$

同理,由微分中值定理对任意 j

$$g_j(x_k + td_k) = g_j(x_k) + t\nabla g_j(x_k + \xi_j d_k)d_k, \quad \xi_j \in [0, t].$$

如果 $j \notin J(x_k)$,那么由 $g_j(x_k) < 0$ 可知存在充分小的 $t_j > 0$ 使得

$$g_j(x_k + td_k) \leqslant 0, \quad \forall t \in [0, t_j];$$

如果 $j \in J(x_k)$,则由引理5.3.4知 $\nabla g_j(x_k)^T d_k < 0$,从而存在充分小的 t_j 使得

$$g_j(x_k + td_k) \leqslant 0, \quad \forall t \in [0, t_j].$$

令 $\bar{t} = \min\{t_0, t_1, \cdots, t_m\}$,则对任意 $t \in [0, \bar{t}]$,(5.2.7)式成立. 证毕.

引理 5.3.6 设 x^* 是由算法产生的点列 $\{x_k\}$ 的一个聚点并且设 $\{x_k\}_K$ 是收敛到 x^* 的子列. 则 $\{d_k^0\}_K \rightarrow 0$, $\{\lambda_k\}_K \rightarrow \lambda^* \geqslant 0$,这里 λ^* 是对应于 x^* 的 Lagrangian 乘子. 亦即,x^* 是问题(5.1.1)的 KKT 点.

证明 由 (5.2.5)式和 (5.2.7)式及引理 5.3.4 知,序列 $\{f(x_k)\}$ 是下降的. 根据假设 5.2.1,有

$$f(x_k) \rightarrow f(x^*), \quad k \xrightarrow{K} \infty.$$

又因为 $\nabla f(x_k)^T d_k^0 - \lambda_k^T g_{J_k} = -(d_k^0)^T H_k^{-1} d_k^0$,所以有

$$0 \leftarrow f(x_{k+1}) - f(x_k) \leqslant \begin{cases} -\alpha \, (d_k^0)^T H_k^{-1} d_k^0, & \text{if } x_{k+1} = x_k + d_k^0 \\ \alpha \, t_k \, \nabla f(x_k)^T d_k^1, & \text{否则} \end{cases}$$

$$\text{(5.3.2)}$$

如果 $x_{k+1} = x_k + d_k^0$,根据假设 5.3.1 和算法的 Step 3,可得

$$\{d_k^0\} \xrightarrow{K} (d^0)^* = 0, \qquad \{\lambda_k\} \xrightarrow{K} \lambda^* \geqslant 0.$$

这就是说 x^* 是一个问题 (5.1.1)的 KKT 点.

如果 $x_{k+1} = x_k + t_k d_k$,若 $d_k^1 \xrightarrow{K} (d^1)^* = 0$,则根据引理 5.3.2 结论是正确的. 用反证法. 假设 $d_k^1 \xrightarrow{K} (d^1)^* \neq 0$,那么存在常数 $d' > 0$ 使得对充分大的 $k \in K$ 有 $\| d_k^1 \| \geqslant d'$. 下面,我们要证存在 $t' > 0$ 使得对所有充分大的 $k \in K$ 有 $t_k \geqslant t'$. 因而,$t_k d_k^1 \nrightarrow 0$.

因为 $d_k^1 \rightarrow (d^1)^* \neq 0$,所以存在 $\sigma_1 > 0$, $\rho' > 0$ 使得对充分大的 k 有

$$\nabla f(x_k)^T d_k \leqslant \theta \, \nabla f(x_k)^T d_k^1 \leqslant -\sigma_1 < 0, \quad \rho_k > \rho' > 0.$$

如果 $j \notin J_0(x^*)$(这里 $J_0(x) = \{j \in I: g_j(x) = 0\}$),那么 $g_j(x_k) \rightarrow$

$g_j(x^*) < 0.$ 因此,存在 $\sigma_2 > 0$ 使得对充分大的 k,

$$g_j(x_k) \leqslant -\sigma_2 < 0.$$

如果 $j \in J_0(x^*)$,那么根据(5.3.1)式,由 $g_j(x_k) \to g_j(x^*) = 0$ 可得

$$\nabla g_j(x_k)^T d_k^1 \to u_{*j} \leqslant 0, \qquad k \xrightarrow{K} \infty$$

进一步又有

$$\nabla g_j(x_k)^T d_k^2 \to -\parallel (d^1)^* \parallel \leqslant -d' < 0, \qquad k \xrightarrow{K} \infty.$$

故对充分大的 k 有

$$\nabla g_j(x_k)^T d_k \leqslant -\frac{1}{2}\rho' d' < 0.$$

由此不难证明存在 $t' > 0$ 使得对充分大的 k 有 $t_k \geqslant t'$ 并且 $t_k d_k^1 \not\to 0$.

另一方面,根据(5.3.2)式和 d_k^1 的定义,可得

$$0 \leftarrow \nabla f(x_k)^T d_k^1 = -(d_k^0)^T H_k^{-1} d_k^0 - \sum_{\lambda_{kj}^1 < 0}(\lambda_{kj}^1)^2 - \sum_{\lambda_{kj}^1 \geqslant 0}\lambda_{kj}^1 \mid g_j \mid \leqslant 0$$

因此,易见有下式成立

$$(d^0)^* = 0, \lambda^* \geqslant 0 \Rightarrow (d^1)^* = 0.$$

此与假设 $(d^1)^* \neq 0$ 矛盾. 证毕.

根据引理 5.3.6,立即可得下面的算法收敛性定理.

定理 5.3.1 算法或有限步终止于问题 (5.1.1)的 KKT 点,或产生一无穷点列,其任一聚点都是问题(5.1.1)的 KKT 点.

§5.4 算法的收敛速度

在这一节里,讨论算法的收敛速度. 我们用下面的假设 5.4.1 替代原

来的假设 5.2.1,并另外增加两个假设. 它们在本节一直成立.

假设 5.4.1 对任意 $j \in I$, $f(x)$ 和 $g_j(x)$ 具有二阶连续偏导数;

假设 5.4.2 强二阶充分条件成立,即,

$$d^{\mathrm{T}} \nabla^2_{xx} L(x^*, \lambda^*) d > 0, \qquad \forall d \in \ker \nabla g_{\hat{J}(x^*)}(x^*) \backslash \{0\},$$

这里 $L(x, \lambda) = f(x) + \lambda^{\mathrm{T}} g(x)$, $g(x) = (g_1(x), \cdots, g_m(x))^{\mathrm{T}}$, $\hat{J}(x^*) = \{j \in J(x^*) : (\lambda^*)_j > 0\}$, 而 (x^*, λ^*) 是问题(5.1.1)的 KKT 对;

假设 5.4.3 $\| (H_k - \nabla^2_{xx} L(x^*, \lambda^*)) d_k^0 \| = o(\| d_k^0 \|)$.

引理 5.4.1 设 (x^*, λ^*) 是问题 (5.1.1) 的 KKT 对. 则存在一个 (x^*, λ^*) 的凸邻域 Ω 和一个正数 μ 使得对所有的 $(x_k, \lambda_k) \in \Omega$, 矩阵

$$\hat{M}_k = \begin{bmatrix} \nabla^2 L(x_k, \lambda_k) & A_k \\ A_k^{\mathrm{T}} & 0 \end{bmatrix}$$

非奇异并且有 $\| \hat{M}_k^{-1} \| \leqslant \mu$.

证明 见 [11, Proposition 3.1].

引理 5.4.2 若算法没有在有限步终止,则在假设 5.4.1、假设 5.4.2 和假设 5.4.3 下, k 充分大后算法将不再进入 Step 6.

证明 显然,如果 k 充分大后,一直有 $\| d_k^0 \| \leqslant \delta$, $\lambda_{kj} \geqslant -\eta \| d_k^0 \|$ 成立. 那么根据假设 5.4.1,定理 5.3.3(i) 和 (QP),只要 $\| d_k^0 \|$ 足够小,(5.2.5)式就总是成立的. 这说明算法在迭代过程中将不再进入 Step 6,而一直在 Step 3 进行迭代使得 $x_{k+1} = x_k + d_k^0$. 故我们只需证 k 充分大后 $\| d_k^0 \| \leqslant \delta$, $\lambda_{kj} \geqslant -\eta \| d_k^0 \|$ 成立.

如果结论不真,那么算法将进入 Step 6 无限多次. 由于 (d_k^0, λ_k) 是子问题(QP)的 KKT 对,所以下面的等式成立

$$\left[\begin{bmatrix} H_k^{-1} - \nabla^2 L(x_k, \lambda_k) & 0 \\ 0 & 0 \end{bmatrix} + \hat{M}_k \right] \begin{bmatrix} d_k^0 \\ \lambda_{J_k} \end{bmatrix} = \begin{bmatrix} -\nabla f(x_k) \\ -g_{J_k}(x_k) \end{bmatrix}.$$

根据假设 5.4.3，由上式可得

$$\hat{M}_k \begin{bmatrix} d_k^0 \\ \lambda_{J_k} \end{bmatrix} = \begin{bmatrix} -\nabla f(x_k) \\ -g_{J_k}(x_k) \end{bmatrix} + o(\parallel d_k^0 \parallel),$$

并且进一步有

$$\hat{M}_k \begin{bmatrix} x_k + d_k^0 - x^* \\ \lambda_{J_k} - \lambda_{J_k}^* \end{bmatrix}$$

$$= \begin{bmatrix} -\nabla f(x_k) + \nabla^2 L(x_k, \lambda_k)(x_k - x^*) - A_k \lambda_{J_k}^* \\ -g_{J_k}(x_k) + A_k^{\mathrm{T}}(x_k - x^*) \end{bmatrix} + o(\parallel d_k^0 \parallel).$$

设 $(x_k, \lambda_k) \in \Omega$. 由假设 5.4.1，$\nabla L(x^*, \lambda^*) = 0$ 和微分中值定理，我们有

$$\parallel -\nabla f(x_k) + \nabla^2 L(x_k, \lambda_k)(x_k - x^*) - A_k \lambda_{J_k}^* \parallel$$

$$\leqslant \parallel \nabla f(x^*) - \nabla f(x_k) + \nabla^2 f(x_k)(x_k - x^*) \parallel + \parallel \sum_{J_k} \lambda_j^* \nabla g_j(x^*) +$$

$$\sum_{J_k} \lambda_{kj} \nabla g_j^2(x_k)(x_k - x^*) - \sum_{J_k} \lambda_j^* \nabla g_j(x_k) \parallel$$

$$\leqslant \parallel \nabla f(x^*) - \nabla f(x_k) + \nabla^2 f(x_k)(x_k - x^*) \parallel + \parallel \sum_{J_k} (\lambda^*)_j [\nabla g_j(x^*) -$$

$$\nabla g_j(x_k) + \nabla^2 g_j(x_k)(x_k - x^*)] \parallel +$$

$$\parallel \sum_{J_k} [\lambda_{kj} - (\lambda^*)_j] \nabla^2 g_j(x_k)(x_k - x^*) \parallel$$

$$\leqslant \parallel \int_0^1 [\nabla^2 f(x_k + t(x^* - x_k)) - \nabla^2 f(x_k)](x_k - x^*) \mathrm{d}t \parallel +$$

$$c_1 \sum_{J_k} \parallel \int_0^1 [\nabla^2 g_j(x_k + t(x^* - x_k)) - \nabla^2 g_j(x_k)](x_k - x^*) \mathrm{d}t \parallel +$$

$$c_2 \sum_{J_k} \| \lambda_{kj} - (\lambda^*)_j \| \cdot \| x_k - x^* \| = o(\| x_k - x^* \|)$$

其中 c_1, c_2 是常数.

同理,由 $J_k \subseteq J(x^*)$, $g_{J_k}(x^*) = 0$ 和假设 5.4.1 可得

$$\| -g_{J_k}(x_k) + A_k^T(x_k - x^*) \|$$

$$= \| g_{J_k}(x^*) - g_{J_k}(x_k) + A_k^T(x_k - x^*) \|$$

$$\leqslant \sum_{J_k} \| g_j(x^*) - g_j(x_k) + \nabla g_j(x_k)^T(x_k - x^*) \|$$

$$= o(\| x_k - x^* \|).$$

这样,不管 x_k 是由 Step 3 还是 Step 6 确定的,如果记 $z_{k+1} = x_k + d_k^0$,由于 $\hat{J}(x^*) \subseteq J_k$,总有

$$\| z_{k+1} - x^* \| = o(\| x_k - x^* \|) \tag{5.4.1}$$

$$\| \lambda_{J_k} - \lambda_{J_k}^* \| = o(\| x_k - x^* \|) \tag{5.4.2}$$

由于算法的收敛性,即 $d_k^0 \to 0$ $(k \to \infty)$ 知,必有充分大的 k,使得 $\| d_k^0 \| \leqslant \delta$ 即 x_{k+1} 是由 Step 3 确定的. 那么,由(5.4.1)式得

$$\| d_k^0 \| = \| x_{k+1} - x_k \| \geqslant \| x_k - x^* \| -$$

$$\| x_{k+1} - x^* \| \geqslant \frac{1}{2} \| x_k - x^* \|.$$

即,

$$\| x_k - x^* \| \leqslant 2\delta.$$

因此,

$$\| d_{k+1}^0 \| = \| z_{k+2} - x_{k+1} \| \leqslant \| z_{k+2} - x^* \| + \| x_{k+1} - x^* \|$$

$$\leqslant 2 \| x_{k+1} - x^* \| \leqslant \frac{1}{4} \| x_k - x^* \| \leqslant \frac{1}{2} \delta$$

又根据(5.4.2)式，对 $j \in J_{k+1}$ 有

$$\lambda_{k+1,j} \geqslant \lambda_{k+1,j} - (\lambda^*)_j \geqslant - |\lambda_{k+1,j} - (\lambda^*)_j|$$

$$\geqslant - \|\lambda_{k+1} - \lambda^*\| \geqslant - \frac{\eta}{2} \|x_{k+1} - x^*\|$$

$$\geqslant - \eta \|z_{k+2} - x_{k+1}\| = -\eta \|d_{k+1}^0\|.$$

上面两个不等式说明第 $(k+2)$ 次迭代将在 Step 3 进行并且 $x_{k+2} = x_{k+1} + d_{k+1}^0$. 因此从 $k+1$ 开始，迭代过程将一直在 Step 3 中进行. 矛盾. 证毕.

由引理 5.4.2 知，k 充分大后算法将在牛顿步中迭代，所以下列定理成立.

定理 5.4.1 在满足本节所有的假设下算法是超线性收敛的，即 k 充分大后有

$$\|x_{k+1} - x^*\| = o(\|x^k - x^*\|).$$

§5.5 数值试验

本节给出算法的数值试验结果. 四个算例选自 [57]，用 MATLAB 编程计算. 每一步迭代，H_k 由 BFGS 公式修正. 参数取值 $\alpha = 0.3, \theta = 0.75, \beta = \eta = 0.5$.

算例 5.5.1 (Modified Rosen-Suzuki problem, No. 264 of [57]):

$$\min f(x) = x_1^2 + x_2^2 + 2x_3^2 + x_4^2 - 5x_1 - 5x_2 - 21x_3 + 7x_4$$

$$s.t. \ g_1(x) = x_1^2 + x_2^2 + x_3^2 + x_4^2 + x_1 - x_2 + x_3 - x_4 - 8 \leqslant 0$$

$$g_2(x) = x_1^2 + 2x_2^2 + x_3^2 + 2x_4^2 - x_1 - x_4 - 9 \leqslant 0$$

$$g_3(x) = 2x_1^2 + x_2^2 + x_3^2 + 2x_1 - x_2 - x_4 - 5 \leqslant 0$$

算例 5.5.1 的最优解和最优值分别是 $x^* = (0, 1, 2, -1)$ 和

$f(x^*) = -44.$ 初始点取为 $x_0 = (-2, 1, -1, 1)$. 当 $k = 35$ 时,得

$$x_k = (-0.006\,678\,48,\ 0.995\,368\,07,\ 2.002\,864\,89,\ -0.998\,969\,89),$$

$$f(x_k) = -43.984\,721\,30.$$

算例 5.5.2 (No. 268 of [57]). 设

$$D = \begin{bmatrix} -74 & 80 & 18 & -11 & -4 \\ 14 & -69 & 21 & 28 & 0 \\ 66 & -72 & -5 & 7 & 1 \\ -12 & 66 & -30 & -23 & 3 \\ 3 & 8 & -7 & -4 & 1 \\ 4 & -12 & 4 & 4 & 0 \end{bmatrix}, \quad d = \begin{bmatrix} 51 \\ -61 \\ -56 \\ 69 \\ 10 \\ -12 \end{bmatrix}.$$

考虑下列问题:

$$\min f(x) = x^{\mathrm{T}} D^{\mathrm{T}} Dx - 2d^{\mathrm{T}} Dx + d^{\mathrm{T}} d$$

$$s.t.\ g_1(x) = x_1 + x_2 + x_3 + x_4 + x_5 - 5 \leqslant 0$$

$$g_2(x) = -10x_1 - 10x_2 + 3x_3 - 5x_4 - 4x_5 + 20 \leqslant 0$$

$$g_3(x) = 8x_1 - x_2 + 2x_3 + 5x_4 - 3x_5 - 40 \leqslant 0$$

$$g_4(x) = -8x_1 + x_2 - 2x_3 - 5x_4 + 3x_5 + 11 \leqslant 0$$

$$g_5(x) = 4x_1 + 2x_2 - 3x_3 + 5x_4 - x_5 - 30 \leqslant 0.$$

算例 5.5.2 的最优解是 $x^* = (1, 2, -1, 3, -4)$,最优值是 $f(x^*) = 0$. 初始点取为 $x_0 = (1, 1, 1, 1, 1)$. 当 $k = 25$ 时,得

$$x_k = (0.992\,726\,08,\ 1.991\,851\,08,\ -0.997\,372\,60,$$

$$2.981\,532\,00,\ -3.965\,596\,81),$$

$$f(x_k) = 4.333\,917\,6 \times 10^{-5}.$$

算例 5.5.3 (No. 269 of [57]).

$$\min f(x) = (x_1 - x_2)^2 + (x_2 + x_3 - 2)^2 + (x_4 - 1)^2 + (x_5 - 1)^2$$

$$s.\,t.\ \ x_1 + 3x_2 = 0,$$

$$x_3 + x_4 - 2x_5 = 0,$$

$$x_2 - x_5 = 0.$$

算例 5.5.3 的最优解是 $x^* = (-0.767\,4,\ 0.255\,8,\ 0.627\,9,$ $-0.116\,3,\ 0.255\,8)$，最优值是 $f(x^*) = 4.093\,02$. 取初始点 $x_0 = (2,2,2,2,2)$. 经过 15 次迭代，即 $k = 15$，得到

$$x_k = (-0.767\,412,\ 0.255\,798,\ 0.627\,901,\ -0.116\,315,\ 0.255\,797),$$

$$f(x_k) = 4.093\,022\,1.$$

算例 5.5.4 (No. 285 of [57]). 问题为

$$\min f(x) = C^{\mathrm{T}} x$$

$$s.\,t.\ \sum_{j=1}^{15} a_{ij} x_j^2 - b_i \leqslant 0, \quad i = 1,2,\cdots,10.$$

其中 C 是 15 维向量，$A = \{a_{ij}\}$ 是 10×15 阶矩阵，b 是 10 维向量. 具体数值参见文献[57].

算例 5.5.4 的最优解是 $x^* = (1,1,\cdots,1)'$，最优值是 $f(x^*) = -8\,252$. 取初始点 $x_0 = (0,0,\cdots,0)$，当 $k = 40$ 时得到

$$x_k = (1.038\,7,\ 0.961\,63,\ 0.976\,95,\ 0.988\,87,\ 0.974\,49,$$

$$1.047\,7,\ 0.857\,01,\ 0.979\,57,\ 1.014\,6,\ 1.052\,4,$$

$$1.012\,7,\ 0.999\,2,\ 0.973\,53,\ 1.013\,4,\ 1.005)$$

$$f(x_k) = -8\,245.9$$

参 考 文 献

[1] Bazaraa M. S. , Sherali H. D. , Shetty C. M.. Nonlinear Programming: Theory and Algorithms (Second Edition). New York: John Wiley & Sons, Inc 1993.

[2] Beck A. and Teboulle M.. Global Optimality Conditions for Quadratic Optimization Problems with Binary Constraints. SIAM J. Optim. 2000, 11(1): 179 - 188.

[3] Bell D. E. and Shapiro J. F.. A Convergent Duality Theory for Integer Programming. Operations Research, 1977, 25: 419 - 343.

[4] Ben-tal A. and Zibulebsky M.. Penalty/ Barrier Multiplier Methods for Convex Programming Problems. SIAM J. Optim. 1997, 7(2): 347 - 366.

[5] Bonnans J. F. and Panier E. R. etc.. Avoiding the Maratos Effect by Means of a Nonmonotone Line Search. SIAM J. Numer. Anal. , 1992, 29(4): 1187 - 1202.

[6] Bonnans J. F.. Local Analysis of Newton Type Methods for Variational Inequalities and Nonlinear Programming. Applied Math. Optim. 1994, 29: 161 - 186.

[7] Bonnans J. F. and Launay G.. Sequential Quadratic Programming with Penalization of the Displacement. SIAM J. Optimization, 1995, 54(4): 792 - 812.

[8] Cetin, B. C. Barhen, J. and Burdick, J. W.. Terminal Repeller Unconstrained Subenergy Tunneling for Fast Global Optimization, JOTA, 1993, 77(1): 97 - 125.

[9] Di Pillo. G. and Lucidi S. On Exact Augmented Lagrangian Functions in Nonlinear Programming, Nonlinear Optimization and Applications. Edited by G. Di Pillo and F. Giannessi, New York: Plenum Press, 1996, 85 – 124.

[10] Dixon, L. C. W. and Szegö, G. P. (Eds). Towards Global Optimization. North-Holland, Amsterdam, 1975.

[11] Facchinei F. and Lucidi S.. Quadratically and Superlinearly Convergent for the Solution of Inequality Constrained Minimization Problems. JOTA, 1995, 85: 2, 265 – 289.

[12] Fang S. C. and Puthenpura S.. Linear Optimization and Extensions (Theory and Algorithm). Prentice Hall, Inc. 1993.

[13] Fisher M. L.. The Lagrangian Relaxation Method for Solving Integer Programming Problems. Management Science, 1981, 27: 1 – 18.

[14] Forsgren A. , Gill P. E. and Wright M. H.. Interior Method for Nonlinear Optimization. SIAM Review, 2002, 44 (4): 525 – 597.

[15] Gao Z. Y. , He G. P. and Wu F.. Algorithm of Sequential System of Linear Equation for Nonlinear Optimization Problem, part I — Inequality Constrained Problems, Technical Report 94 – 31. Inst. of Applied Math. , Chinese Academy of Sciences, China 1994.

[16] Ge R. P.. The Theory of Filled Function Methods for Finding Global Minimizers of Nonlinearly Constrained Minimization Problems. J. of Comput. Math. , 1987, 5(1): 1 – 9.

[17] Ge, R. P. and Qin, Y. F.. A Class of Filled Functions for Finding a Global Minimizer of a Function of Several

Variables. Journal of Optimization Theory and Applications，1987，54(2)：241－252.

[18] Ge，R. P.. A Filled Function Method for Finding a Global Minimizer of a Function of Several Variables. Math. Programming，1990，46：191－204.

[19] Ge Renpu. Finding More and More Solutions of a System of Nonlinear Equtions. Applied Math. and Computation，1990，36：15－30.

[20] Ge，R. P. and Qin，Y. F.. The Globally Convexized Filled Functions for Global Optimization. Applied Math. and Computation，1990，35：131－158.

[21] Geoffirion A. M.. Lagrangian Relaxation for Integer Programming. Math. Programming Stud，1974，2：82 -114.

[22] Goh C. J. and Yang X. Q.. A Sufficient and Necessary Conditin for Nonconvex Constrained Optimization. Appl. Math. Lett. ，1997，10：9－12.

[23] Gonzaga C. C. and Castillo R. A.. A Nonlinear Programming Algorithm Based on Non-coercive Penalty Functions. Math. Program，Ser. A，2003，96：87－101.

[24] Guignard M. and Kim S.. Lagrangian Decomposition: A Model Yielding Stronger Lagrangian Relaxation Bounds. Mathematical Programming，1993，33：262－273.

[25] Han，Q. M. and Han，J. Y.. Revised Filled Function Methods for Global Optimization. Applied Math. and Computation，2001，119：217－228.

[26] Han S. P.. Superlinearly Conergence Variable Metric Algorithms for General Nonlinear Programming Problems. Mathematical Programming，1976，11：263－282.

[27] Holland J. H. Adaptation in Natural and Artificial Systems.

MIT Press, 1975.

[28] Horst R. A New Branch and Bound Approach for Concave Minimization Problems. Lecture Notes in Computer Science, 1976, 41: 330 - 337.

[29] Horst R. An Algorithm for Nonconvex Programming Problems. Mathematical Programming, 1976, 10: 312 - 321.

[30] Horst R. A General Class of Branch-and-Bound Methods in Global Optimization with Some New Approaches for Concave Minimization. Journal of Optimization Theory and Applications, 1986, 51: 271 - 291.

[31] Horst R. Deterministic Methods in Contrained Global Optimization: Some Recent Advances and New Fields of Application. Naval Research Logistics, 1990, 37: 433 - 471.

[32] Horst, R. , Pardalos, P. M. and Thoai, N. V.. Introdution to Global Optimization. Kluwer Academic Publishers, Dordrecht, Netherland, 1995.

[33] Horst R. , Pardolos P. M. (Eds.) Handbook of Global Optimization. Dordreht, The Netherlands, Kluwer Academic Publishers, 1995.

[34] Horst R. , Tuy H. Global Optimization (Deterministic Approaches), 3rd ed. berlin, Germany: Springer, 1994.

[35] Huang H. Y.. Uniformly Approach to Quadratically Convergent Algorithm for Function Minimization. JOTA, 1970, 5.

[36] Huyer W. , Neumaier A.. A New Exact Penalty Function. SIAM J. OPTIM. 2003, 13(4): 1141 - 1158.

[37] Karmarkar N.. A New Polynomial-time Algorithm for Linear Programming. Combinatorica, 1984, 4: 373 - 395.

[38] Kirkpatrick S. , Gelatt C. D. , and Vecchi M. P..

Optimization by Simulated Annealing. Science，1983，220：671－680.

[39] Kleimmichel H. and Schoneleld K.. Newton-type Methods for Nonlinearly Constrained Programms, Proceedings of the 20th Jahrestagung "Mathematische Optimierung". Humboldt-Universitat. Zu Burlin, Seminarberichte, 1988, 53－57.

[40] Kleinmichel H., Richetr C. and Schonefeld K.. On a Class of Hybrid Methods for Smooth Constrained Optimizatin. JOTA, 1992, 73：3, 465－499.

[41] Konno H., Thach P. T., Tuy H. Optimization on Low Rank Nonconvex Structures. The Netherlands：Kluwer Academic Dordrecht, 1997.

[42] Levy, A. V. and Montalvo, A.. The Tunneling Algorithm for the Global Minimization of Functions. SIAM Journal on Scientific and Statistical Computing, 1985, 6(1)：15－29.

[43] Li D.. Zero Duality Gap in Integer Programming: p-Norm Surrogate Constraint Method. Operations Research Letter, 1999, 25：89－96.

[44] Liu, X.. Finding Global Minima with a Computable Filled Function. Journal of Global Optimization, 2001, 19：151－161.

[45] Llewellyn D. C. and Ryan J.. A Primal Dual Integer Programming Algorithm. Discrete Appl. Math, 1993, 45：262－273.

[46] Lucidi, S. and Piccialli, V.. New Classes of Globally Convexized Filled Functions for Global Optimization. Journal of Global Optimization, 2002, 24：219－236.

[47] Lundy M. and Mess A. Convergence of an Annealing Algorithm. Math. Prog. 1986, 34：111－124.

[48] Maratos N.. Exact Penalty Function Algorithm for Finite Dimensional and Contral Optimization Problems. Ph. D. Thesis, University of London, UK 1978.

[49] Mayne D. Q. and Polak E.. A Superlinearly Convergent Algorithm for Constrained Optimization Problems. Math. Programming Study, 1982, 16: 45 - 61.

[50] Michelon, P. N. and Maculan, N.. Lagrangian Decomposition for Integer Nonlinear Programming with Linear Constrains. Mathematical Programming, 1991, 52: 303 - 313.

[51] Ng C. K. , Li D. and Zhang L. S.. Global Descent Method for Global Optimization. The Chinese University of Hong Kong, Ph. D. thesis, 2003.

[52] Oblow E. M. A Stochastic Tunneling Algorithm for Global Optimization. JOGO, 2001, 20: 195 - 212.

[53] Peng J. M. and Yuan Y. X.. Optimality Conditions for the Minimization of a Quadratic with Two Quadratic Constraints. SIAM J. Optim. 1997, 7(3): 579 - 594.

[54] Polyak R.. Modified Barrier Functions (theory and mathod). Mathematical Programming, 1992, 54: 177 - 222.

[55] Powell M. J. D.. The Convergence of Variable Metric Method for Nonlinear Constrained Optimization Calculations. Nonlinear Programming 3, New York: Academic Press, 1978, 27 - 63.

[56] Rockafellar R. T.. Convex Analysis. Prentice Hall, 1970.

[57] Schittkowski K.. More Test Examples for Nonlinear Programming Codes. Springer-Verlag, 1987.

[58] Shapiro J. F.. A Survey of Lagrangian Techniques for Discrete Optimization. Annals of Discrete Mathematics,

1979, 5: 113 - 138.

[59] Stephens C. P. and Baritompa W.. Global Optimization Requires Global Information. Journal of Optimization Theory and Applications, 1998, 96(3): 575 - 588.

[60] Strekalovsky A. S.. Global Optimality Conditions for Nonconvex Optimization. J. of Global Optimization, 1998, 12: 415 - 434.

[61] Tuy H. , Thieu T. V. , Thai N. Q. A Conical Algorithm for Globally Minimizing a Concave Function over a Closed Convex Set. Mathematics of Operations Research, 1985, 10: 498 - 514.

[62] Tuy H. , Khachaturov V. , Utkin S. A Class of Exhaustive Cone Splitting Procedures in Conical Algorithms for Concave Minimization. Optimization, 1987, 18: 791 - 807.

[63] Tuy H. , Horst R. Convergence and Restart in Branch — and — Bound Algorithms for Global Optimization Application to Concave Minimization and D. C. Optimization Problems. Mathematical Programming, 1989, 41: 161 - 183.

[64] Tuy H.. Normal Sets, Polyblocks and Monotone Optimization. Vietnam J. Math. , 1999, 27: 277 - 300.

[65] Tuy H.. Monotone Optimization: Problems and Solution Approaches. SIAM J. Optim. , 2000, 11(2): 464 - 494.

[66] Wang Wei. A Superlinearly Convergent Algorithm by the Suboptimal Method for Linear Constrained Covex Programming Problems. J. Sys. Sci. & Math. Scis. , 1990, 10(1): 31 - 39.

[67] Wang W. and Xu Y. F.. A Fitting Method for Unconstrained Minimization. OR Transaction, 2003, 7(1): 46 - 52.

[68] Wang W. and Xu Y. F.. An Asymptotic Strong Duality

Method for Integer Programming. Acta mathematica scientia, 2004, 24: 285 - 292.

[69] White D. J.. Weighting Factor Extensions for Finite Multiple Objective Vector Minimization Problems. European Journal of Operations Researchn, 1988, 36: 256 - 265.

[70] Williams H. P.. Duality in Mathematics and Linear and Integer Programming. JOTA, 1996, 90: 257 - 278.

[71] Wolpert D. H. , Macready W. G. No Free Lunch Theorems for Search. Santa Fe Institute: Technical Report SFI-TR-95 -02 - 010, 1995.

[72] Wu F. and Guei X. Y.. A Kind of Variable Metric Algorithm with n+1 Parameters. ACTA Mathematica Sinica (in Chinese), 1981, 24: 6 - 16.

[73] Wu F. A Superlinear Converget Projection Type Algorithm for Linearly Constrained Problems. European Journal of Operational Research, 1984, 16: 334 - 344.

[74] Wu Z. Y. , Bai F. S. , Yang X. Q. Zhang L. S.. An Exact Lower Order Penalty Function and its Smoothing in Nonliear Programming. Optimization, 2004, 53(1): 51 - 68.

[75] Xu Y. F.. Quasi-Newton Methods: Analysis and Algorithm, Ph. D. Dissertation. Institute of Applied Mathematics, Chinese Academy of Sciences, 1998.

[76] Xu Y. F. and Wang W.. A Mixed Superlinearly Convergent Algorithm with Nonmontone Search for Constrained Optimizations. Appl. Math. J. Chinese Univ. 2000, 15(2): 211 - 213.

[77] Xu Y. F. and Wang W.. A QP-Free and Superlinearly Convergent Algorithm for Inequality Constrained Optimizations. Acta Mathematica Scientia, 2001, 21B(1): 121 - 130.

[78] Xu Y. F. and Wang W.. A Feasible Algorithm for Inequality Constrained Optimization by Means of Nonmonotone Linear Search. OR Transactions, 2001, 5(1): 1-12.

[79] Xu Y. F., Wang W. & Gao Z. Y.. The Algorithm of Sequential KKT Equations by Nonmonotone Search for Arbitrary Initial Point. Computational Optimization and Applications, 2001, 18: 221-232.

[80] Xu Z., Huang H. X., Pardalos P. M. etc. (2001). Filled Functions for Unconsreained Global Optimization. Journal of Global Optimization, 20, 49-65.

[81] Yao Y.. Dynamic Tunneling Algorithm for Global Optimization. IEEE Transactions on Systems, Man and Cybernetics, 1989, 19(5): 1222-1230.

[82] Yuan Y.. On a Subproblem of Trust Region Algorithm for Constrsined Optimization. Math. Programming, 1990, 47: 53-63.

[83] Zhang J. L. and Wang C. Y.. A New Conjugate Projection Gradient Method. OR Transaction, 1999, 3(2): 61-70.

[84] Zhang L. S.. A Sufficient and Necessary Condition for a Globally Exact Penalty Function. Chinese Journal of Contemporary Mathematics, 1997, 18(4): 415-424.

[85] Zhang L. S., Li D.. Global Search in Nonliear Integer Programming: Filled Function Approach. International Conference on Optimization Techniques and Applications, Perth, 1998, 446-452.

[86] Zhang L. S., Gao F., and Zhu W. X.. Nonlinear Integer Programming and Global Optimization. J. of Computational Mathematic, 1999, 17(2): 179-190.

[87] Zhang Q. and Zhang L. S.. Global Minimization of

Constrained Problems with Discontinuous Penalty Functions. Computers and Mathematics with Applications，1999，37：41－58.

[88] Zhang L. S., Ng C. K., Li D. and Tian W. W.. A New Filled Function Method for Global Optimization. Journal of Global Optimization，2004，28：17－43.

[89] 邬冬华.求全局优化的积分型算法的一些研究和新进展.上海大学博士论文，2002.

[90] 邬冬华，田蔚文，张连生.一个求总极值的实现算法及其收敛性.运筹学学报，1999,3(2)：82－89.

[91] 邬冬华，田蔚文，张连生等.一种修正的求总极值的积分—水平集方法的实现算法收敛性.应用数学学报，2001,24(1)：100－110.

[92] 吴至友.全局优化的几种确定性方法.上海大学博士论文，2003.

[93] 席少霖.非线性最优化方法.北京：高等教育出版社,1992.

[94] 袁亚湘，孙文瑜.最优化理论与方法.北京：科学出版社,2001.

[95] 张连生.L_1精确罚函数和约束总极值问题.高校计算数学学报,1988，2：141－148.

[96] 赵瑞安，吴方.非线性最优化理论和方法.杭州：浙江科学技术出版社,1992.

作者攻读博士学位期间
发表和已投稿的论文

[1] Xu Y. F., Wang W. & Gao Z. Y. The Algorithm of Sequential KKT Equations by Nonmonotone Search for Arbitrary Initial Point. Computational Optimization and Applications, 2001, 18: 221-232.

[2] Xu Y. F., Wang W. A QP-free and Superlinearly Convergent Algorithm for Inequality Constrained Optimizations. Acta Mathematica Scientia, 2001, 21B(1): 121-130.

[3] Xu Y. F., Wang W. A Feasible Algorithm for Inequality Constrained Optimization by Means of Nonmonotone Line Search. OR Transactions, 2001, 5(1): 1-12.

[4] Wang W., Zhang L. S. A Revised Conjugate Projection Gradient Algorithm for Inequality Constrained Optimizations. ICMP2002..

[5] Wang W., Xu Y. F. A Rank-one Updating Algorithm for Unconstrained Optimization Problems. ICMP2002.

[6] Wang W., Xu Y. F. A Fitting Method for Unconstrained Minimization. OR Transaction, 2003, 7(1): 46-52.

[7] 罗雪梅，王薇，韩跃军. 并发事务无死锁的可串行化调度的形式化方法. 计算机工程与应用, 2004, 40 (10): 181-185.

[8] 罗雪梅，王薇，王朝阳. 数据之间相互转换的并行算法. 山东科技大学学报, 2003, 22(4): 49-51.

[9] 王薇，徐以汎. 整数规划的渐进强对偶方法. 数学物理学报 (A), 2004, 24: 285-292.

[10] Wang W., Zhang L. S., Xu Y. F. A Conjugate Projection Gradient Method for Inequality Constrained Optimization Problems. submit to Journal of Computational Mathematics.

[11] Wang W., Yang Y. J., Zhang L. S. Unification of Filled Functions and Tunnelling Functions for Global Optimizations. submit to Acta Mathematicae Appl.

致　谢

值此论文完成之际,我谨向我的导师张连生教授表示衷心的感谢和崇高的敬意.我的工作得益于张老师的精心指导和耐心鼓励.无论在学术研究还是教书授业中张老师严肃认真的态度,锐意进取的精神,严谨治学的作风和循循善诱的方法都给我留下了深刻印象,使我受益匪浅.他严于律己,宽以待人亦为我们树立了榜样.

由衷感谢我的家人对我的爱心、理解和无私奉献.

感谢同济大学数学系领导为我的学习提供方便和支持.

非常感谢田蔚文教授对我的热情帮助.

论文第二章的数值计算是在师弟杨永建的帮助下完成的,谨表谢意.

感谢同门兄弟姐妹几年来对我的鼓励和友谊.

感谢所有在我学业和工作上与我合作,给予我帮助的人们.